MW00967150

The Handbook of Machine Soldering

Second Edition

The Handbook of Machine Soldering

A Guide for the Soldering of Electronic Printed Wiring Assemblies

Second Edition

Ralph W. Woodgate
President
WoodCorp. (The Woodgate Corporation)

A Wiley-Interscience Publication
JOHN WILEY & SONS
New York • Chichester • Brisbane • Toronto • Singapore

Library of Congress Cataloging-in-Publication Data

Woodgate, Ralph W., 1922-

The handbook of machine soldering : a guide for the soldering of electronic printed wiring assemblies / Ralph W. Woodgate. -- 2nd ed.

p. cm.

"A Wiley-Interscience publication."

Includes index.

ISBN 0-471-85779-3

1. Solder and soldering. 2. Printed circuits. I. Title.

TT267.W66 1988 88-6076

621.381'74--dc 19 CIP

Printed in the United States of America

10 9 8 7 6 5 4 3 2

To Catherine and Samantha
for all the days spent at the typewriter

Foreword

We are fortunate that users of soldering machinery can now have a better reference text that covers this technology at a layman's level but with enough theory to provide a good education on the subject. It is also nice to see that some of the mysteries of this field dealing with the natural phenomena of capillary attraction have been unlocked. The information on equipment is very enlightening and noncommercial.

This volume is a very strong, informative text on a most meaningful subject.

JIM D. RABY, P.E.

Soldering Technology
Naval Weapons Center
China Lake, California

Preface

Mass soldering either by wave, drag, or dip machines has been the preferred method of making high-quality, reliable connections for many decades. In spite of the appearance of new connecting systems, it still retains this position. Correctly controlled, it is one of the lowest-cost methods of making electrical connections. Incorrectly controlled, it can be one of the most costly processes: not because of the initial cost but because of the many far-reaching effects of work incorrectly done.

Until the first edition of this book there was no definitive text available which dealt with the subject in a clear and practical way, and the success of the first edition prompted this revision. The fundamental facts stay the same of course, but I have tried to expand some sections and have added new material. The changes are not dramatic or extensive but are based on conversations with people on the shop floor during my work as a consultant, with many companies both large and small, manufacturing a wide range of electronic products.

I would like to thank all those who contributed in this way and to the many who took the time during the WoodCorp Soldering Workshops to discuss the use of this book. You all helped materially to keep the subject matter fresh and practically useful.

Because I am directly involved in this industry, I have tried to keep commercialism out of the text and to present the facts in an unbiased manner. I have no affiliation with any company but WoodCorp Inc, which is dedicated to providing its clients with impartial advice. However, I would not be fair to my readers if I did not point out some of the strengths and weaknesses of the performance of various types of equipment as I see them function on the shop floor. I have procured information from many commercial sources for this revision, and I thank them all for their assistance. To the reader I say, "Try all that the market place has to offer. Let the results determine the equipment that you buy. Every product and process has slightly different requirements and the only secret is to find the equipment and materials which are the best for your particular purpose."

The data at the back of the book are provided as a service only, and the presence or absence of any company name must not be construed as a comment on the company or its products.

I have directed this book to the people working on the shop floor who are involved in the selection, installation, and operation of the soldering process. It is a very practical book and does not require any particular understanding of the chemistry or physics of soldering. Theoretical purists may be somewhat upset by some of the simplification of the theory but every effort has been made to ensure technical accuracy while avoiding complex mathematical or scientific terms. This book is for all levels of practitioners who have to make the process work.

For too long soldering has been seen as part science, part magic, and never completely predictable. In this book I hope that I will show that this attitude is total nonsense. Soldering is quite a simple process and if correctly controlled is capable of totally repeatable results. It requires care and attention to detail but when these factors are applied with a clear understanding of the process then the results will more than pay for the effort. The elimination of rework, the reduction of cost, and the improvement in product reliability are all factors that will make our industry profitable and ensure continued growth.

RALPH W. WOODGATE

Brewster, New York
September 1987

Contents

The Handbook of Machine Soldering

Second Edition

Chapter 1

The History of Machine Soldering and a Brief Review of Other Soldering Methods

Soldering is not a new process. It probably goes back to the Bronze Age when some metalworker discovered the affinity of a tin-lead mixture for a clean copper surface. Certainly, the Romans used a 60/40 tin-lead mixture to solder their lead water pipes, as recorded in the writings of Pliny. Many examples of this form of soldering are still in existence.

During the late nineteenth and early twentieth centuries, a great deal of work was carried out on the development and improvment of the soldering process, and in determining the fundamental principles involved. An examination of the books, papers, and especially the patents of that era shows that most of the knowledge of soldering was clearly understood a long time before the electronics industry began. There has certainly been a great improvement in the materials and machines used in the process, but the basics have changed little since the early part of this century.

It was not until the development of the printed wiring board, usually known as a PWB, that it was feasible to develop any form of automated soldering system for the electronics industry. The PWB placed all the connections on one plane, and the modern soldering system as we know it became possible (Fig. 1.1). Initially the boards were hand dipped into a solder bath. With skilled operators it was possible to achieve quite acceptable results with the open packaging and wide circuits of those days. It was inevitable that some faulty joints, shorts, bridges, and other defects were found, but these were easily repaired with the traditional soldering iron. This of course was the beginning of the solder joint inspector and the touch-up operator.

From this beginning the printed wiring board industry and the machine soldering industry have grown together—each dependent on the other for their tremendous rate of growth—and have almost gained universal acceptance as the most economical method of packaging and interconnecting electronic circuitry.

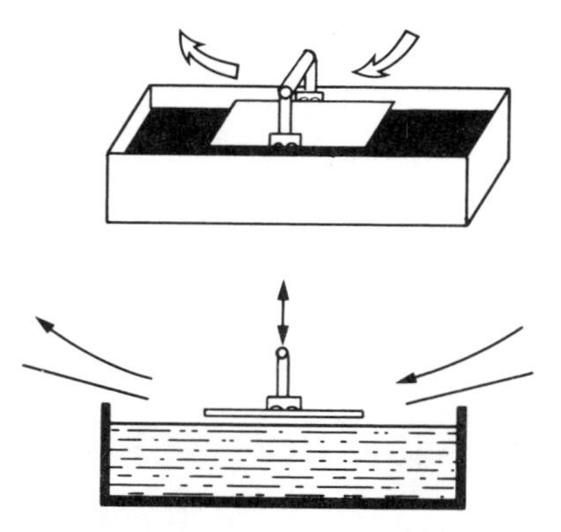

Figure 1.1. Manual solder dipping.

Machine or mass soldering can make thousands of joints in a few seconds, providing the electrical connections and simultaneously mechanically fastening the components. No other process has yet been developed that can so economically interconnect electronic circuitry.

From the original concept of hand dipping boards came the idea of pumping solder through a slot or nozzle to form a wave, or mound of constantly moving, dross-free metal through which the board could be passed. This had several advantages: the surface of the solder was always clean and did not need skimming before soldering, the pumping action maintained an even temperature, and the raised wave provided an easy form for adding a conveyor system for automatic movement of the board over the solder. Once the idea of the solder wave had been born, it was logical to use a similar method to apply the flux, and the basic solder machine appeared. These original machines were quite crude, consisting of individual modules with a freestanding conveyor placed over them, with the conveyor often being added by the user. There was little attempt to integrate the parts into one system, and the smoke and fumes were left to escape into the factory exhaust, aided by a "homemade" canopy hung over the machine.

At this time there was a lot of work done on the development of the actual wave, in an attempt to improve the speed of soldering and to reduce the incidence of shorts and icicles. The solder weir, the double wave, and the cascade all had brief popularity. The cascade wave, which consisted of several waves in series, was one of the more interesting of these developments. It was used with an inclined conveyor and was the basis of some machines built for in-house use by some of the larger companies; however, it never achieved great popularity. Another development at that time was the use of ultrasonics to form a wave, with the intention of causing cavitation at the joint and avoiding the use of flux, or at least reducing the activity required. Some experimental machines were built, but the cost was excessive, the soldering results no better than those produced with the conventional pumped wave, and the power of the ultrasonic energy had to be so high that the matching horns used to energize the pot were quickly eroded away. With the advent of the semiconductor, ultrasonics were found to cause damage to the devices on some occasions, and today the chief use for ultrasonics in soldering in the industry is to be found in the pots used for tinning components prior to assembly. This era also saw the introduction of the use of oil in the solder wave. This was one of the ideas that worked extremely well and has been a feature of some solder machines for many years (Fig. 1.2).

The oil was injected into the wave in several different ways, but the basic idea was the same. The oil on the surface of the wave reduced the surface tension of the solder, assisted in the wetting of the parts to be joined, and aided the excess solder to drain as the board moved out of the wave. Thus, the use of the oil reduced the incidence of shorts, tended to avoid icicles, and reduced the amount of solder used. It also had some effects that were not so advantageous, and the whole question of "oil or no oil"—the so-called *dry wave*—will be discussed in detail in Chapter 5. There is no doubt that at that time the use of oil was a major step forward in the development of the machine soldering process.

For some years, a wave was considered to be the parabolic shape generated when solder is pumped through a plain slot. In more recent times, there have been many improvermnts in the basic wave design, but without doubt the major advance has been the introduction of the "adjustable asymmetrical wave," with an adjustable dam to enable the exit portion of the wave to

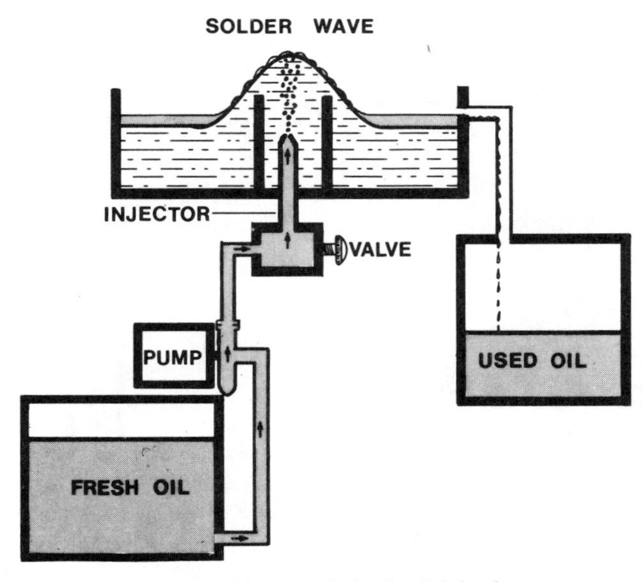

Figure 1.2. Diagram of simple oil injection.

be set to give an almost stationary solder lake. The development of these more sophisticated waves has shown that the wave shape can have a marked effect on the performance of the soldering machine, and indeed that the modern high-density boards, with very tightly packaged components, can be processed much more efficiently using these improved wave shapes. The design of solder waves for specific purposes is now well established as an engineering tool in the solder machine industry (Fig. 1.3).

In a totally different direction the static solder bath soldering process was also developed, and machines were designed using this principle, for example, in the so-called *drag soldering* process (Fig. 1.4). These machines have been developed to a high level of sophistication, initially in Europe and Japan, now in the United States. As has often been the case in the growth of the soldering machine industry, development of this initially started because of patent restrictions that at the time prevented the free use of the more conventional solder wave. These restrictions have often proved beneficial to the industry by forcing creative thinking that has eventually resulted in improved products.

So today there are machines made by many different manufacturers each with its own particular features. Generally, they are all robust and well suited to the task. They have various levels of sophistication and capability, ranging from the small laboratory machine to the high production system, which is often complete with in-line cleaning and assembly and return conveyors. Machines these days are generally self-contained, with a high degree of control and optional auxiliary systems. They are usually complete with exhaust venting. Coming onto the market are systems with computer control of the operation and setup, which practically eliminates the human error from the machine soldering process. In the following chapters the various types and options available will be reviewed so that the correct selection can be made for any particular application (Fig. 1.5).

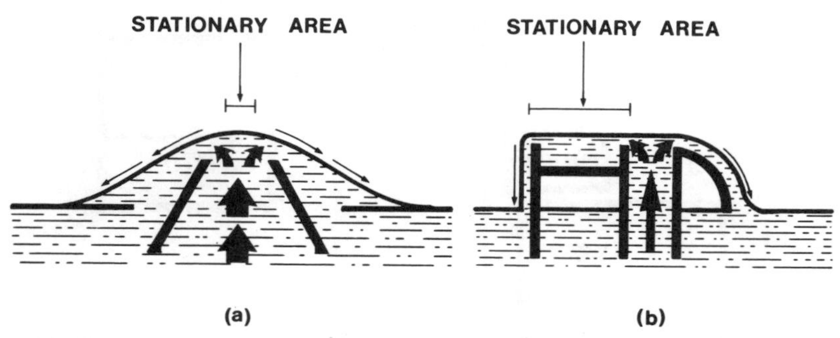

Figure 1.3. Symmetrical and asymmetrical waves. (a) Symmetrical wave. Solder flows equally in both directions: small stationary area. (b) Asymmetrical wave. Most solder flows in one direction: large stationary area.

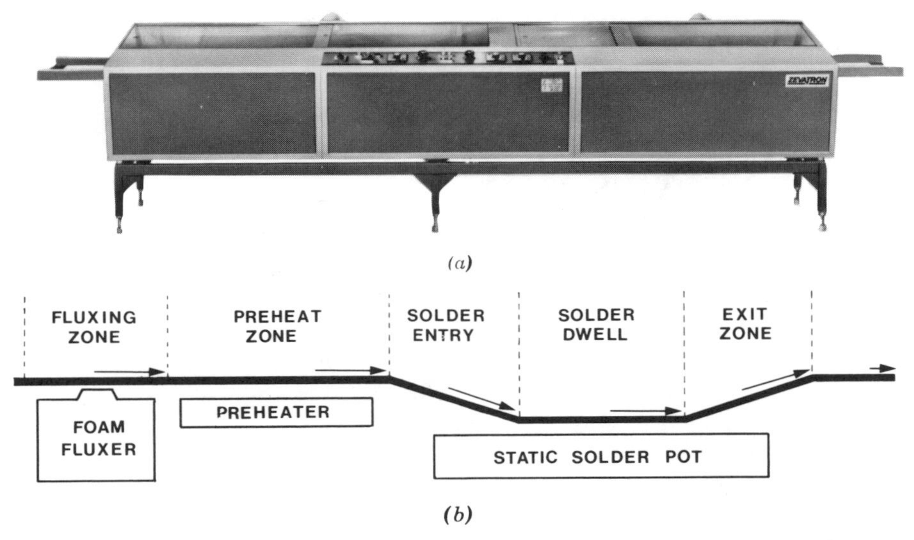

Figure 1.4. Drag soldering. (*a*) Modern drag soldering machine. Courtesy Zevatron GMBH. (*b*) Schematic diagram of drag soldering machine.

In spite of all the improvements in the design of mass soldering systems and machines, it is generally one of the least carefully controlled processes and rarely is its full potential achieved. One measure of this is the ready acceptance of the almost universal "touch-up" operation. When machine soldering was first introduced, deficiencies in the process made a certain amount of rework inevitable. The tremendous improvement in productivity over hand soldering meant that the retention of a few operators for repair and rework was not unreasonable. Today, however, the machines, solder, and fluxes have been so improved that in a properly controlled soldering operation touch-up should be the exception, not the rule.

A great deal of the book concerns the methods and procedures necessary to achieve this excellence in soldering.

Throughout this book the soldering processes described involve the simultaneous application of heat and solder by immersing the joint structure in a wave or pool of molten solder. There are other methods of soldering in which the solder and heat are applied separately. These other processes are not discussed here in detail because they are not part of the subject matter of this book, but it is valuable to know of them and some of the circumstances in which they can be used to supplement the mass soldering process.

Solder can be applied in several ways.

1. A mixture of solder powder and flux, known as *solder paste,* can be applied to the joints by screening, stenciling, or using one of the machines that apply dots of the solder paste by a simple pneumatic system.

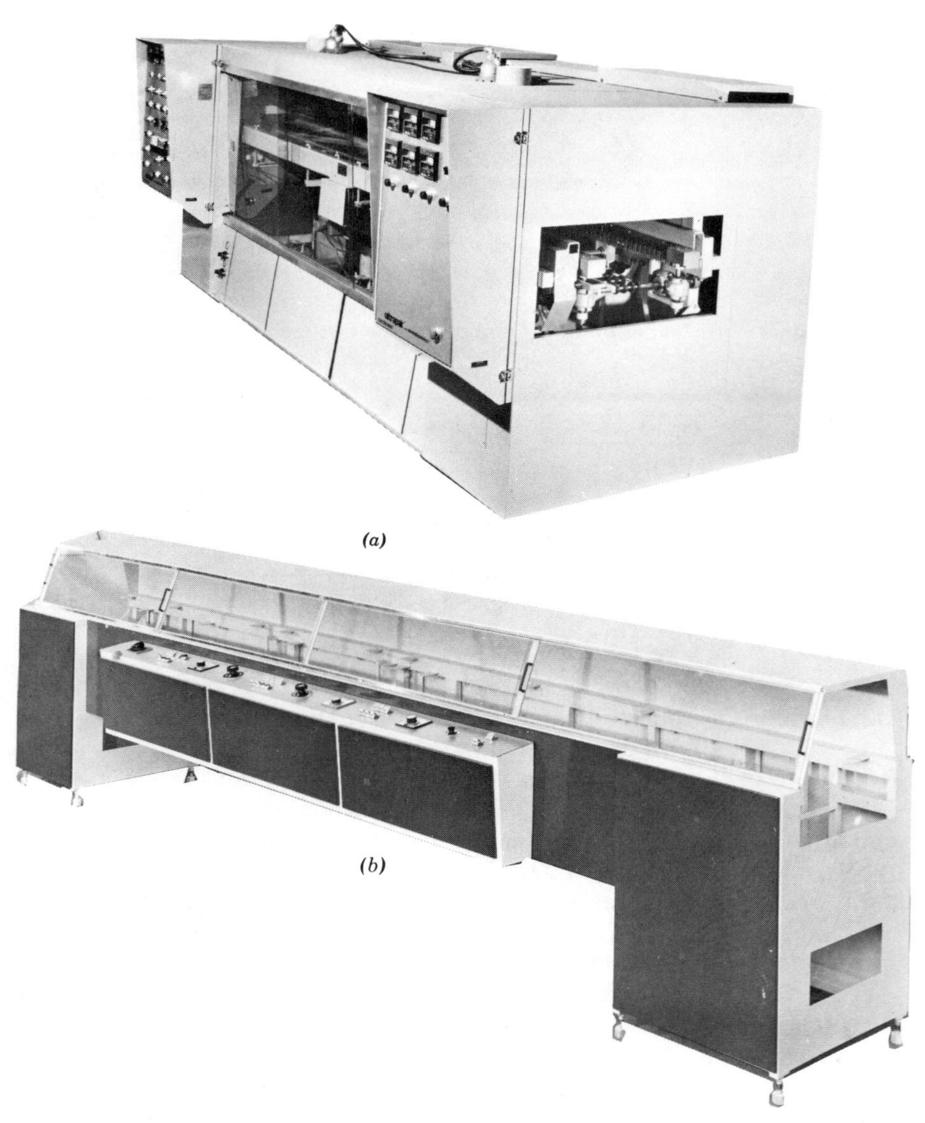

(a)

(b)

Figure 1.5. Typical modern soldering machines. (*a*) Courtesy Electrovert Ltd. (*b*) Courtesy Zevatron GMBH.

(c)

(d)

Figure 1.5. (*c*) Courtesy Light Soldering Development Ltd. (*d*) Courtesy Hollis Engineering.

This method is especially useful where surface mounted components are to be soldered, and they cannot be passed through the solder wave.

2. The parts can be pretinned, that is, coated with a layer of solder by fluxing and dipping into a bath of the molten metal.
3. Solder can be obtained in the shape of rings, washers, or tubes, which are placed on or adjacent to the parts to be joined. These solder *preforms,* as they are called, can be obtained with or without a flux coating.

These two latter methods of providing solder for the joint are used in many areas of the electronics industry where machine soldering is not possible. Tiny solder rings are placed over long wire wrap pins on a mother board where the gold plating of the pins in the wrapping area must not be solder coated. Solder washers are fitted under bolts that must be soldered to a ground plane. Wires are tinned before soldering into a pretinned wiring lug. The possibilities are limitless, and more will quickly come to mind.

Applying the solder is only half of the task of making the soldered joint; it now has to be heated. There are many obvious ways of doing this, such as oven heating or using a blow torch or a hot soldering iron. There are also several less well-known ways of providing the heat necessary to melt the solder and complete the joint.

1. Condensation or vapor phase soldering is becoming widely used. In this process the vapor from a high temperature boiling point liquid is allowed to condense on the items to be soldered and, in giving up the latent heat of vaporization, quickly and evenly raises their temperature and produces a sound soldered joint. The method is simple, clean, and extremely well controlled. It requires special equipment and, of course, the parts are all subject to the same temperature.
2. The parts can be dipped into an oil heated above the melting point of the solder. Peanut oil can be used, although there are some excellent synthetic products available for this purpose. This is a messy process, and here again the component parts are subject to the same temperature as the joint and will require cleaning after soldering to remove the oil.
3. Infrared heating is a useful method of producing the temperature necessary for soldering. The shorter wavelengths can be focused into very small areas, allowing selective heating of the joints to be soldered, without subjecting the components to soldering temperature. The equipment and tooling is expensive if complex assemblies have to be soldered, and the relative emissivity of the joint metals and the adjacent materials can cause processing problems. For example, the base laminate of a PWB can be burned while the joint to be soldered has not reached soldering temperature. This process therefore requires very careful control.
4. Resistance heating can be used for soldering. It is done in two ways. The first method passes a current through the parts of the joint, causing their

temperature to rise by virtue of the resistance existing in the joint structure. This is not a very controllable process and is chiefly used for less critical applications on large connections. The second method passes a current through a wire that is shaped to touch the parts to be heated. Again this is not an easy process to control precisely, but it is quite often used to solder the leads on surface mounted components.

Lasers are used to provide the heat for soldering, although it is not yet a common method. The laser can provide a very high-energy, small-area beam, and heating is therefore extremely fast and limited only by the ability of the joint materials to absorb this energy. However, the shiny surfaces of the metals in the joint reflect most of the energy of the commonly used lasers, and coupling the laser and the joint is a major problem. The flux is often used as a coupling medium. When correctly set up, laser soldering offers a fast, clean method of making joints—not without some problems—and the cost of the equipment is quite high.

There are many different methods of applying solder, heat, and, of course, flux to produce a soldered joint. It is almost always possible to develop from these a process that will produce satsfactory joints under any particular circumstance. These methods are so obviously very much less efficient than machine soldering, where many thousands of joints can be made in a few seconds with a well-controlled process that uses comparatively low-cost materials and equipment.

Chapter 2

The Theory

Soldering is a very simple process. The only things necessary to produce a perfect joint are the following.

Solderable parts, correctly configured
The correct temperature for the correct time
The right composition of flux and solder

In spite of this simplicity, soldering sometimes seems to be one of the major problem areas of the electronics industry and a mixture of science, art, and black magic.

The answer, of course, is that the conditions above are often difficult to achieve, especially the solderability requirement. In addition, in the machine soldering of PWBs, the board design can have an important part to play in the efficiency of the soldering process. Too often these factors are either ignored or forgotten, and they cannot be compensated for later on, except by the expenditure of unnecessary labor in inspection and touch-up.

When soldering problems arise, there are always good reasons, inevitably to be found in the areas mentioned above. Provided the fundamental factors are clearly understood and the problems are tackled in a logical manner, there should be no difficulty in arriving at a solution. There is no reason to accept anything but excellence in the soldering process.

These fundamentals will be examined in a practical manner aimed at providing a totally adequate understanding of the processes, without requiring a detailed knowledge of the chemistry and physics involved.

THE SOLDER JOINT

If two strips of clean copper are soldered together, and a section is cut across the joint, the surface of the copper will be easily identified by the

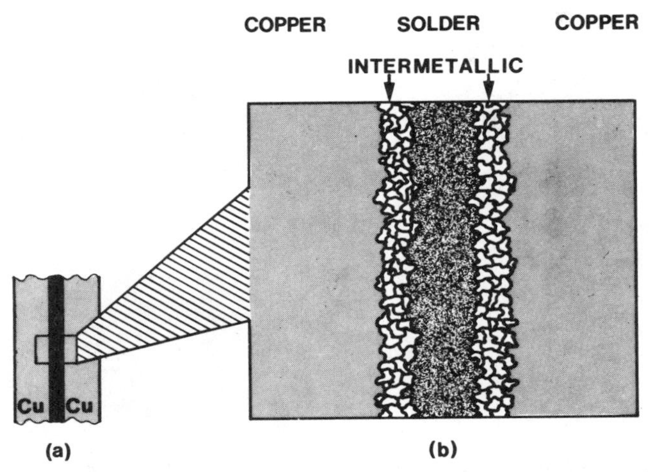

Figure 2.1. The solder joint. (*a*) Two strips of copper soldered together. (*b*) Magnified view of a small section of the joint.

typical color of the metal. The solder will appear as a thin silver line between the two copper parts (Fig. 2.1*a*). If this section is magnified, it will become apparent that the joint contains something more than solder and copper, and what appears to be a thin fine line of solder to the naked eye is in fact a much more uneven and complex affair (Fig. 2.1*b*). Under the microscope the copper surfaces will lie on either side of a mottled mass of solder, and between the copper and the solder there is another metallic formation of a totally different appearance.

This joint structure is the heart of the soldering process, and a clear understanding of the way in which it is formed, and the materials contained in it, is absolutely necessary if the rest of the soldering theory is to make sense and be of practical use.

The mottled appearance of the solder is caused by the separation and clumping together of the crystals of the lead and tin during the cooling of the alloy. The clumps of lead crystals appear darker. The size and shape of these masses of crystals will depend on the cooling rate of the solder joint and the composition of the solder.

The metallic formation between the solder and the copper is obviously something formed during the making of the joint. It was not there prior to the soldering of the copper strips. It is called an *intermetallic compound* (Fig. 2.2) and is an important part of the soldering process. There cannot be a satisfactory copper-solder-copper joint without the formation of this intermetallic alloy. It proves that the solder has wetted or bonded to the copper and become joined to it by an atomic bond. The solder is not acting as some sort of super glue; it has become an integral part of the metal structure of the joint. In fact, the tin in the solder has amalgamated with some of the copper

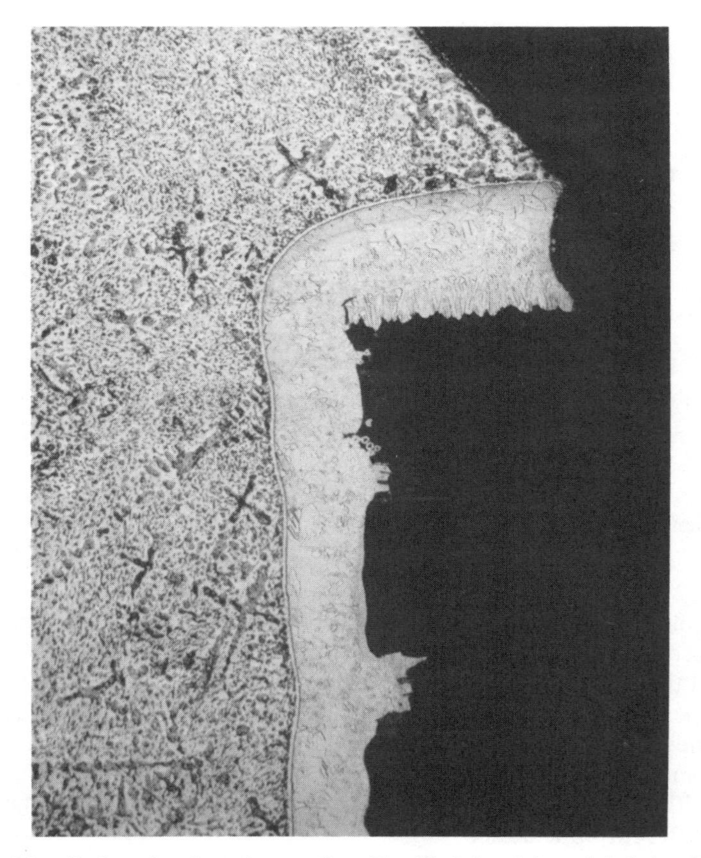

Figure 2.2. Magnified section through part of a solder filled plated through hole in a PWB. The intermetallic compound is clearly visible between the solder and the copper plating.

to form a new alloy. If this alloy is not present then the solder has not "wetted" the copper, and, instead of a true soldered joint with a high mechanical strength and excellent electrical conductivity, there will only be two pieces of copper stuck together with solder.

This will produce a very weak joint and poor or intermittent electrical conductivity. In fact, if there is no wetting the parts will probably not even hold together and will certainly part company under conditions of shock or vibration, or if subjected to variations of temperature. It is obvious that in electrical and electronic equipment, where reliable operation is most important, care must be taken to see that soldered joints are correctly made to avoid a high rate of product failure.

There is considerable evidence to show that the formation of an intermetallic compound is not necessary to ensure a wetted, soldered joint. In practice, when considering copper as the material to be joined, which is the case in most electronic connections, the time-temperature relationship involved in making the joint will inevitably produce this alloy. This intermetallic compound is an important factor when considering the mechanical properties of the joint and will be discussed in some detail. However, it must be recognized that when other materials are soldered an intermetallic compound may not be formed, and if it is, it may not have the importance of that in the copper-solder-copper joint.

When the solder "wets" the copper, the tin in the solder alloy combines with a tiny part of the copper and forms the intermetallic compound. As the solder cools, this mixture of tin and copper solidifies at the junction of the solder and the copper. The tin-copper alloy is harder than either the copper or the solder and effectively forms a very strong bond between the metals. However, in addition to being hard it is also much more brittle than either of the other two metals, and, if the layer of intermetallic alloy is too thick, the reliability of the joint can be in jeopardy from cracking, which will begin in this alloy. This can be a problem especially if the joint is exposed to any mechanical forces, such as the expansion and contraction of the PWB laminate caused by variations in temperature (Fig. 2.3).

Every copper-solder interface will have this intermetallic compound. the parameters that determine its thickness are time and temperature; therefore, soldering should always be carried out as fast as possible and at the lowest temperature that will produce a satisfactory joint. These requirements also indicate a possible jeopardy to joint quality in the touch-up of unsatisfactory joints. The efforts to produce a joint where there is a difficulty with solderability usually results in several attempts with the soldering iron to produce wetting. Even if the joint is eventually made, the chances are that the solderable parts will have a considerable thickness of intermetallic compound. This also indicates the greater reliability of the machine-made joint, where time and temperature are closely controlled and repeatable (Fig. 2.4).

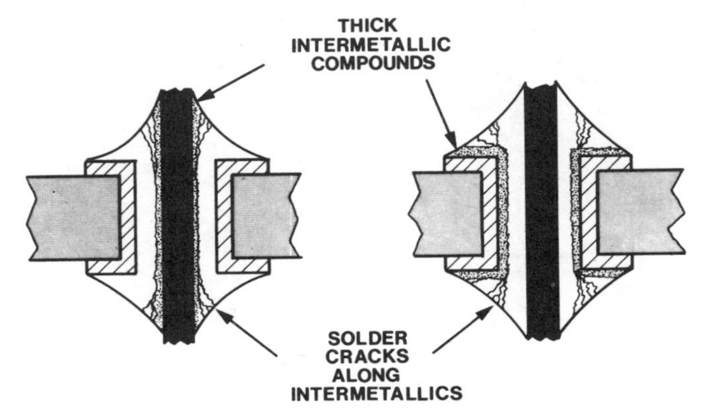

Figure 2.3. Diagrams of typical cross sections of reworked joints showing how cracking can start in a thick, brittle intermetallic compound.

Finally, the time-temperature relationship of the formation of the intermetallic compound suggests that the first joint made on the solder machine is the most reliable, and every precaution must be made to ensure that it is perfect. Touch-up is expensive, unreliable, uncontrollable, and unnecessary. The objective must be to attain zero defect soldering.

Now that the formation and the properties of the intermetallic compound are understood, the possible problems must not be overemphasized. Under normal machine soldering conditions the formation of this compound will not be a problem or even a cause for concern. However, understanding this does explain some of the factors that will be discussed when the setting of the soldering parameter is looked at in detail, and it is the first area to be considered if solder joint cracking is ever a problem. Any concern regarding the thickness of intermetallic compound produced during soldering can be quickly alleviated by making a cross section of the joint and measuring the thickness under the microscope. This technique is discussed in detail in Chapter 7.

An intermetallic compound is of course formed whenever solder wets to a solderable surface, for example, a PWB or a component lead. Once the intermetallic compound is formed it will continue to grow due to the diffusion of copper into the tin of the solder. The rate of diffusion is extremely slow at room temperature, but like most chemical reactions proceeds much faster at elevated temperatures, for example, when PWBs are baked prior to soldering.

If the tin/lead or tin coating is very thin then the tin can become totally converted to compounds of tin and copper. Two different compounds are formed when copper is the base material. While free tin is available, the compound Cu_6Sn_5 will eventually appear at the surface of the coating. As the

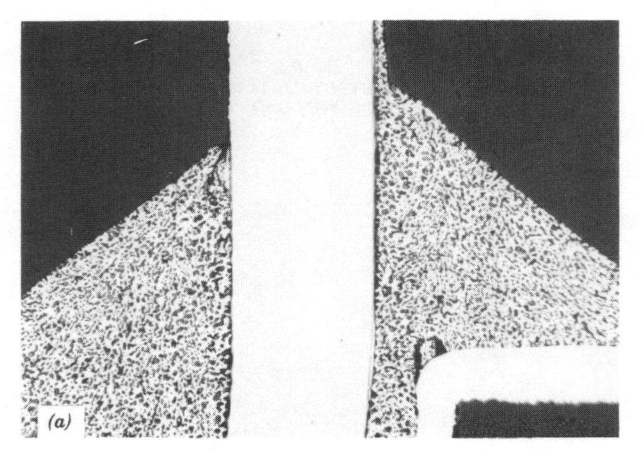

Figure 2.4. (*a*) Cross section of a joint showing cracking propagating along the intermetallic compound between the lead and the solder.

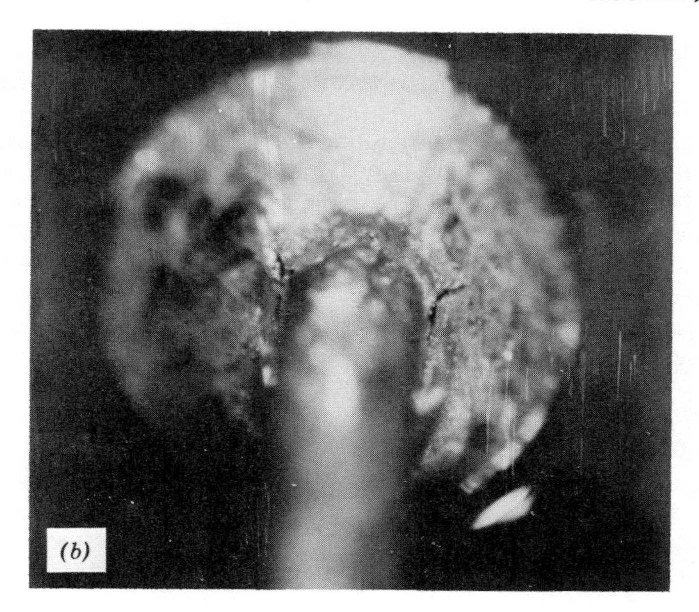

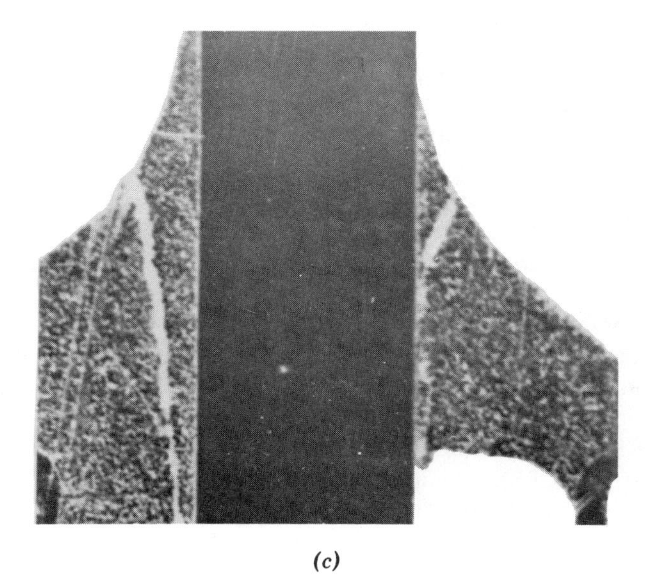

(c)

Figure 2.4. (*b*) Microphotograph of an actual cracked joint. (*c*) Cross section of the cracked solder fillet.

tin is consumed, the compound changes to Cu_3Sn. The former compound, Cu_6Sn_5, cannot be wetted when a rosin based flux is used, while Cu_3Sn cannot be wetted using any of the fluxes normally found in the electronics industry. The formation of these intermetallic compounds is the chief cause of solderability problems, especially in the case of the PWBs where the very thin tin/lead at the knee of the hole in plated through hole (PTH) boards can quite quickly become converted to an intermetallic compound. This causes the phenomenon known as "weak knees," which can prevent the formation of a top side solder fillet. It is obvious therefore that the thickness of the tin or tin/lead solderable coating is extremely important if a long solderable life is to be guaranteed.

WETTING

In order to make a soldered joint the solder has to "wet" the base metal. This concept of wetting with solder is not difficult to understand. If water is poured onto a greasy plate it will ball up and run off, leaving a dry surface (Fig. 2.5). One of the advantages of the more popular car waxes is that the rain beads up and does not wet the body of the automobile. When the wax is washed off by the detergent used at the car wash, or the greasy plate is washed in the dishwasher, the water no longer beads up, but forms a thin film over the surface of the plate or the automobile. It wets the surface, and even when the surplus water is removed, the surface still stays wet. The water has become in some way more closely attached to the object and cannot be simply shaken off (Fig. 2.6).

The only difference in the surfaces is that in one case they were covered with wax, or grease, and in the other they were clean. The clean surface could be wetted, the greasy one could not.

In a similar but more complex manner, solder will not wet the surface of a metal unless it is completely, chemically clean. If the solder is to bond there must be a metal to metal contact between the solder and the base metal. Anything that prevents this contact, for example, grease (even a fingerprint) or corrosion (the thinnest invisible film of oxide) will also prevent the solder from bonding and forming an intermetallic compound. ONLY A PERFECTLY CLEAN SURFACE WILL SOLDER. This fact cannot be emphasized too strongly. More soldering problems are caused by contamination of the parts than by any other cause.

Any surface contaminant, even those that are totally invisible, will cause solder to ball up and not wet the metal, in the same way that the water balled up on the greasy plate. Some may not cause such a drastic reaction, but all can jeopardize the quality of the joint. Cleanliness is one of the keys to excellence in soldering (Fig. 2.7).

When the metal surface is absolutely, chemically clean, not only visibly clean, the solder will wet the surface and flow out as seen in Fig. 2.8. When

Figure 2.5. Water will not wet a greasy plate.

Figure 2.6. A clean plate retains a thin film of water.

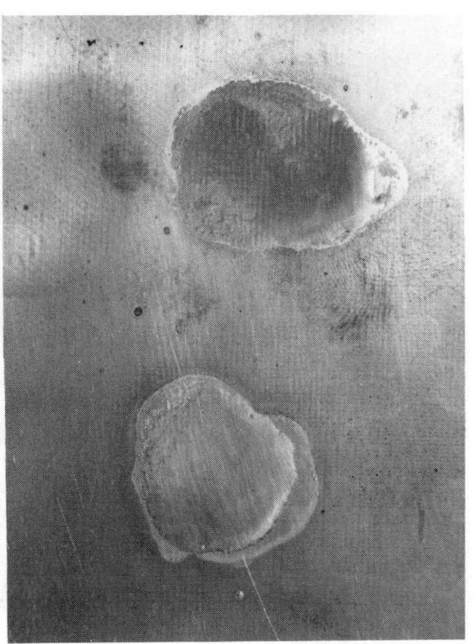

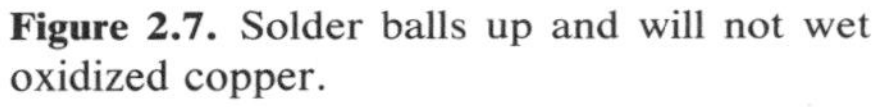

Figure 2.7. Solder balls up and will not wet oxidized copper.

Figure 2.8. Solder flows out and wets a chemically clean copper surface.

the solder has solidified, another difference will be found between the solder that wet the base metal and that which did not. The blob of solder that did not wet can be knocked off easily, leaving little or no solder adhering to the base metal. The solder that flowed out and wet the base metal, on the other hand, will be solidly bonded and can only be removed by scraping or filing. Even if the base metal is heated so that the solder melts, although the surplus can be wiped off, a thin layer will be left, so tightly bonded to the base metal that no amount of wiping can remove it. The solder has become an integral part of the base metal. It has wetted the surface. If the base metal is copper an intermetallic compound has been formed during the wetting process.

Another extremely important difference will be seen between the solder that wetted the base metal and that which did not. This is the angle that the surface of the solder makes to the surface of the base metal (Fig. 2.9). This *wetting angle* is a measure of how well the solder has wetted the metal and is the most important factor in visually judging the effectiveness of the soldering process and the solderability of the base metal. Indeed, it is the most commonly used nondestructive inspection tool and is dealt with at some length in Chapter 7.

Where the solder has wetted the metal it has flowed out and formed a thin feathered edge. Where the solder has not wetted the metal it has a rounded,

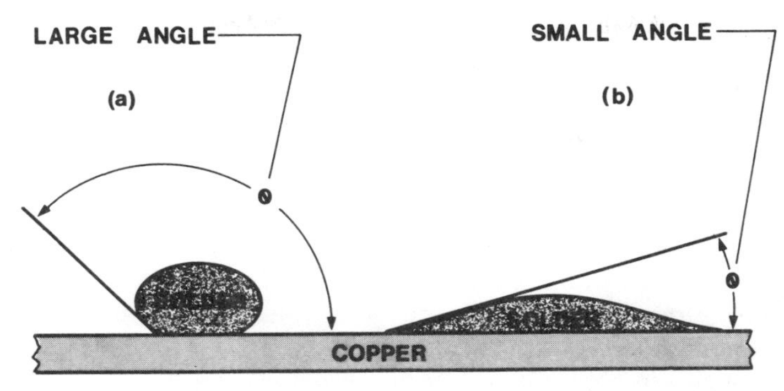

Figure 2.9. The difference in wetting angles between wetted and non wetted solder bonds.

convex shape, and the feathered edge is not present. In practice, variations will be found between these extreme limits, and too much or too little solder in the joint can sometimes cloud the issue. However, with some experience it becomes easy to determine if wetting has occurred and to make an assessment of the quality of the soldered joint. Similarly, by attempting to tin component leads and PWBs, it is possible to use the wetting criterion to check their solderability.

The visual appearance of the soldered joint is usually all that is available to judge its internal quality. It can at best be a judgment; it is usually accurate, but the fact that the internal structure of the joint cannot be seen is another good reason for controlling the materials and the process in order to make this form of subjective inspection unnecessary (Fig. 2.10). Solder in-

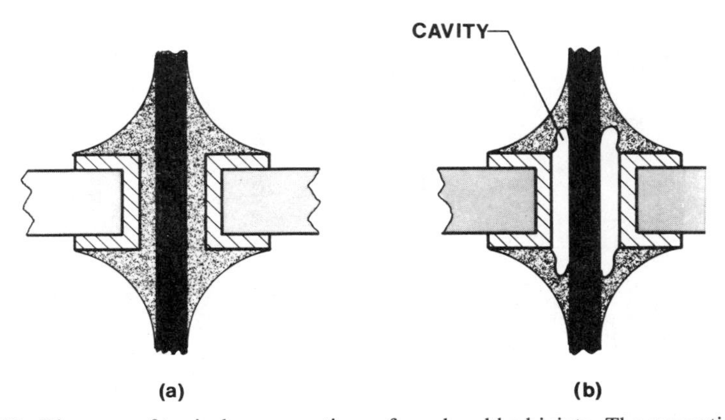

Figure 2.10. Diagrams of typical cross sections of good and bad joints. The cosmetic appearance does not always indicate the quality of the joint.

spection machines of several different basic types are starting to be used. Their different technologies and uses are discussed in Chapter 7.

THE SOLDER JOINT AND THE PRINTED WIRING BOARD

There are three basic forms of solder joint produced in the machine soldering of PWBs: the *single-sided board joint* (Fig. 2.11), the *plated through hole* (PTH) *board joint* (Fig. 2.12), and the *surface mount joint,* which is a form of the single-sided board joint in that the entire mass of the parts to be joined passes through the solder wave. In these figures the illustrations show the magnified cross section taken through the center of the joint. This form of pictorially showing a joint is common in the electronics industry and derives from the cross-sectioning method used to inspect the interior of the joint when troubleshooting any processing problems. In fact, this is the only real method of determining the quality of the joint. However, it is a destructive test and used only in cases of real process failure as a last resort to arrive at a solution.

The illustrations show typical good soldered joints. Note that the solder has wetted the wires and the component mounting pads. It has in the case of the PTH board risen up through the hole, by capillary force, and overflowed, wetting the top surface of the pad, and even risen up the component lead wire. Similarly, in the single-sided board the solder has risen up into the hole slightly and wetted the lead. Sometimes, if the lead has excellent solderability, the solder will completely fill the hole. In all cases the solder has fully wetted the base metal and formed the feather edge on both the component lead and the mounting pad, showing that a good joint has apparently been formed.

In summary:

In order to produce a good soldered joint the solder must wet the base metal.

The solder will only wet the base metal if the surface is perfectly clean.

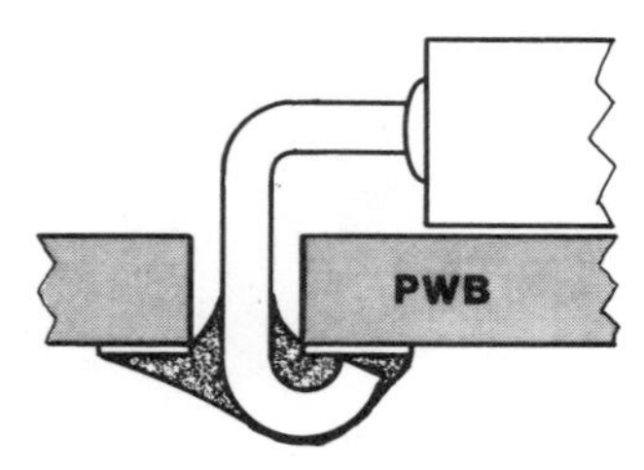

Figure 2.11. A sound solder joint on a single-sided board.

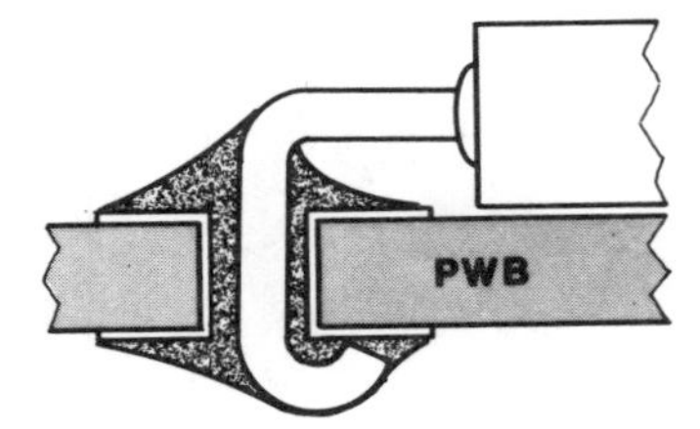

Figure 2.12. A sound soldered joint in a plated through hole board.

When the solder wets the base metal it forms a thin feathered edge and a very small wetting angle.

This wetting angle and the thin feathered edge are the only visual signs of a good joint.

In the copper-solder-copper joint the wetting action will always form an intermetallic compound.

Chapter 3

Fluxes

As discussed in Chapter 2 in the section on Wetting, the solder can only bond to the base metal if the surface is totally clean so that a metal to metal contact can be made. Most metals oxidize in air, so that even if the surface is scraped or filed it will immediately become covered again with a thin layer of oxide that effectively prevents wetting with solder.

This can be demonstrated by scraping copper in an inert atmosphere and applying solder. The solder will wet the copper surface. If a layer of molten solder is placed on a sheet of oxidized copper and the surface scraped through the solder layer, the solder will wet wherever the scraping has removed the oxide and disturbed the solder sufficiently to allow the two metals to come into contact. In both cases the clean metal copper surface has been exposed to the oxide-free solder, and bonding and wetting have occurred.

However, neither of these methods is very practical, and some other way of producing and maintaining a clean surface has to be used. The simple way of achieving these results is to use a flux.

If a ball of solder is placed on a sheet of copper that has been coated with flux (Fig. 3.1*a*) and the copper is placed on a hot plate, then as the copper heats the ball of solder will melt (Fig. 3.1*b*) and flow over the metal surface (Fig. 3.1*c*). The feather edge and the wetting angle both show that the solder has wetted the copper (Fig. 3.1*d*). In close-up, the action that has occurred is shown in Fig. 3.2. As the copper surface becomes hot, the chemical action of the flux dissolves the oxide on the metal and allows the molten solder to wet the chemically clean surface. The solder pushes the hot flux ahead of it as it flows out over the base metal, and as the flux spreads it continues to wet the cleaned surface. This action continues until either the solder moves so far from the heat that it freezes, or there is insufficient solder to continue the flowing action, or there is no more active material available in the flux to continue dissolving the oxide.

With a small amount of solder, even heating, and adequate flux, the rate

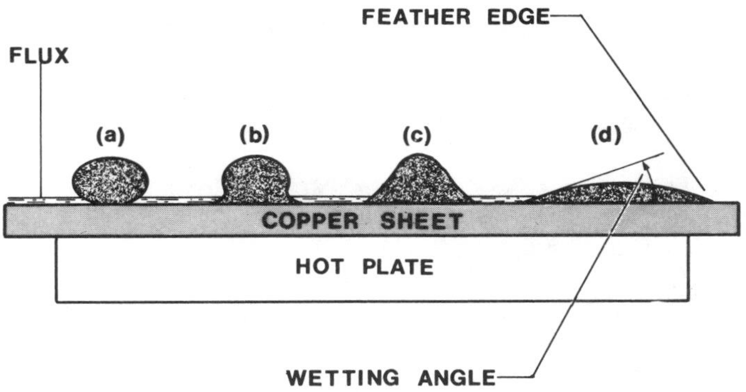

FEATHER EDGE
FLUX
(a) (b) (c) (d)
COPPER SHEET
HOT PLATE
WETTING ANGLE

Figure 3.1. The action of the flux in making a sound soldered joint.

of the solder spreading across the copper and the area wetted are a measure of the effectiveness of the flux. This so-called *spread test* is often used to compare the difference in performance of various fluxes, or as it is called their *activity* (Fig. 3.3). The more active the flux, the faster it will allow the solder to wet the base metal. A highly active flux will permit the soldering of a more heavily oxidized metal. A mildly active flux will only allow wetting of lightly oxidized metals. If a mild flux is used on metals with different amounts of oxidation, the rate and degree of wetting will therefore be a measure of their oxidation and thus their solderability. This principle is used in some of the common solderability tests and is discussed in detail in Chapter 7.

A flux must be sufficiently active to remove the oxides from the metal to be soldered. It must continue to be active at the soldering temperatures and it must be mobile enough to move easily and allow the solder to flow. A flux

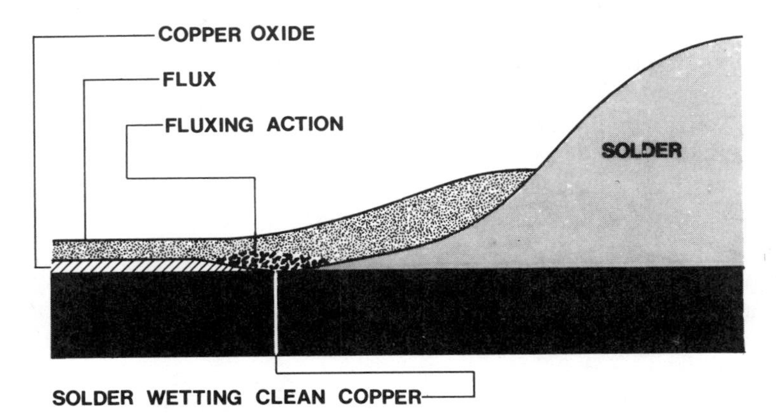

Figure 3.2. A diagram of the fluxing action and solder wetting.

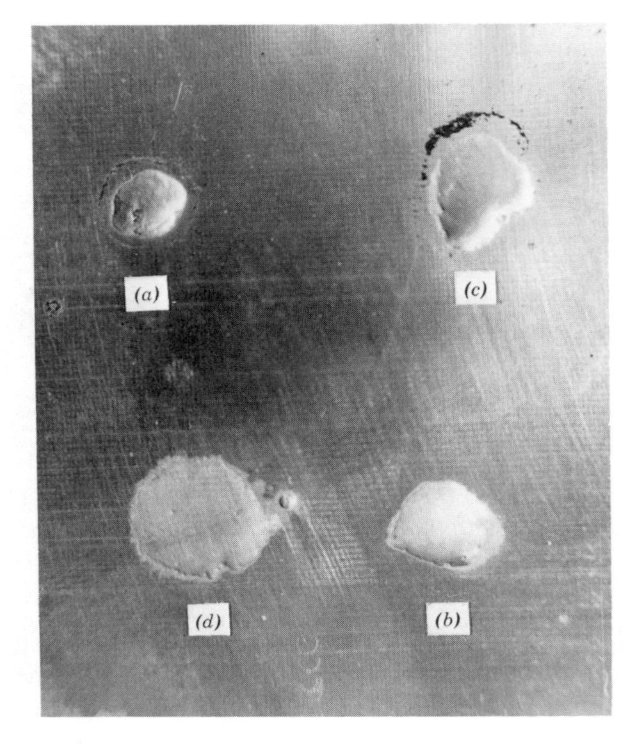

Figure 3.3. The principle of the spread test. (*a*) Plain rosin flux. (*b*) Activated rosin flux. (*c*) Water soluble, organic acid, or intermediate flux. (*d*) Acid flux.

is not, however, a general cleaning agent and must not be expected to remove grease, fingerprints, general dirt, or contamination. It is used only to remove the oxides from the surfaces to be soldered and to prevent them reforming during the soldering process.

There are other properties of fluxes that affect the formation of the joint, for example, the reduction of the surface tension of the solder, some wetting agents to promote solder flow, and antioxidants to avoid flux breakdown at elevated temperatures. These are secondary to the main properties of fluxes and are not part of the basic fluxing action. Perfect soldered connections can be made without these additions, although they assist in making soldering faster, help in reducing solder shorts, and extend the intervals between flux changes.

It would seem logical to use the most active flux available, but unfortunately another factor creeps into the fluxing equation. Because fluxes are able to chemically remove oxides they have to be chemically active; the more active they are the more likely they are to cause corrosion to the components and the PWB and to promote electrical failure. Therefore, the board and components must be designed so that they can be thoroughly cleaned after soldering to remove any flux or flux residues. If this is not possible only the least active fluxes can be used.

SELECTING A FLUX

In many cases the selection of the flux is decided by the product specification, which defines precisely the type of flux permitted. This is especially the case with military or aerospace contracts. However, even when this occurs it is worth trying several different makes of flux because the performance varies from manufacturer to manufacturer even though they all comply with the same specification. When the flux is specified it will be necessary to adjust the soldering parameters, ensure solderability of the parts, and select the cleaning process to accommodate the activity level of the flux to be used. The methods of doing this will become apparent as the book proceeds.

If the choice of flux is open, then there are several factors to be considered in making the flux selection. The dominant concern has nothing to do with the actual soldering, but relates to the cleanliness level required for the completed assembly, the environment in which it will be used, and the need for cleaning. If the assembly has to be cleaned, then the design of the assembly and the components used will decide the cleaning options, and this in turn will decide the types of flux that can be safely used.

For example, in the telephone industry the boards used in the switching and exchange equipment often contain relays that are not hermetically sealed. Therefore, they cannot be immersed in water for cleaning since it would be impossible to rinse and dry them completely, and contamination could remain in the relays and prevent their reliability operation. Solvent cleaning with any of the commonly used materials would similarly leave contaminants on the relay contacts and jeopardize the reliability of the product.

It is common practice, therefore, in this industry to clean only the underside of the soldered assemblies, removing the flux and flux residues so that contact can be made to the soldered joints on the board for automatic testing. Cleaning is accomplished by passing the boards over rotating brushes immersed in a suitable solvent (Fig. 3.4). During fluxing it is inevitable that some flux will pass up through the holes in the board onto the surface of the assembly and will not be removed by the cleaning stage. The boards are used in equipment that has to operate reliably for many years. The flux must corrode neither the boards nor the components and must not affect the insulation between the circuits or the operating conditions. Therefore, the choice of flux is restricted to one of very low activity—inevitably rosin based.

On the other hand, many computer boards contain only DIPs, resistors,

Figure 3.4. Under board brush cleaner. (*a*) Courtesy Electrovert Ltd.

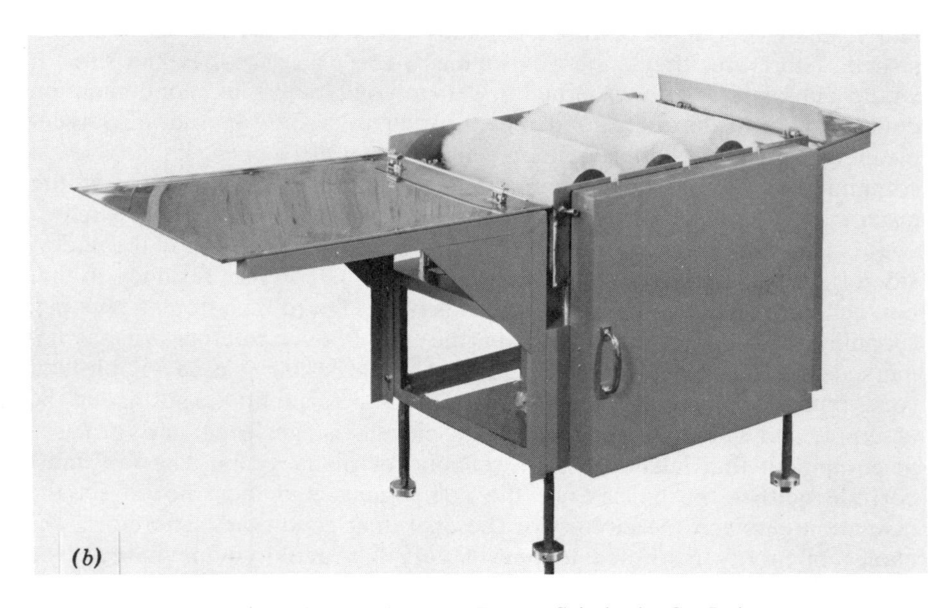

Figure 3.4. (*b*) Courtesy Tamura Seisakusho Co. Ltd.

Figure 3.5. An assembly which is suitable for the use of aqueous or synthetic activated fluxes.

and capacitors, which are all hermetically sealed (Fig. 3.5). These components can also be mounted off the board so that the assembly can be flushed thoroughly during cleaning. Under these circumstances a highly active aqueous flux is an obvious choice. It offers optimum soldering even when the component or board solderability is less then perfect. Immediate cleaning after soldering with only hot water will produce an assembly that is totally free of any flux or flux residues.

A third example is the case of an assembly containing not only sealed components but also a heat sink, transformer, or similar component mounted flush to the board surface. It is inevitable that some flux will find its way under this component by capillary force during fluxing and will be trapped there. It is equally certain that this flux will not be washed out during cleaning. Although the board can otherwise be cleaned perfectly, a noncorrosive, and therefore mildly active, flux must be used to avoid later damage to the flush mounted component and the circuitry running under it. Alternatively, the component can be sealed to the board to prevent the egress of either the flux or the cleaning fluid, and then the choice of flux is not limited.

Whenever the use of a highly active flux is considered, the assembly to be soldered should be examined carefully for places where the flux can be trapped. It is not possible to list all the possible places and components where this can happen, but experience has shown that the following are items to watch out for:

Variable components: potentiometers, trimmers, switches, relays, coils

Insulated components: sleeved capacitors, dipped coils and transformers, ceramic insulated resistors

Sockets and connectors in general

Wires with any form of sleeving or insulation, especially stranded wire and braided screened cables

All these items can be obtained in sealed versions, or the design changed to use other components or methods of connection. In the case of wires and cables it may be possible to solder them by hand after machine soldering the assembly, or they can be sealed against the flux. This usually turns out to be a question of cost, with the expense of the additional labor being compared with the cost savings obtained by using an aqueous flux and water cleaning.

It can be seen from these examples that the selection of a flux is not a subject that can be left until the product is about to go onto the production floor. It must be considered during the design of the package and the specification of the components, and the only way that this will happen is by making the manufacturing engineer an accepted and important member of the design team.

In summary then the selection of a flux is dependent on several factors.

Board design

- Can the board be immersed in solvent or water for cleaning?
- Are the components so mounted that the assembly can be cleaned?

Components selected

- Are they hermetically sealed?
- Can they be cleaned?
- Will they trap residues that may cause later problems?
- Are the components compatible with the cleaning medium?

Product specification

- Will the product see an environment that requires total cleaning?

Within the limitations set by these parameters it makes sense to use the most active flux possible, for the reasons already discussed. Other factors relating to the choice of fluxes will become apparent as the book progresses, the design and use of the various fluxing systems are reviewed, and some of the problems associated with soldering are discussed.

TYPES OF FLUXES

Fluxes for soldering electronic equipment fall into three broad categories, commonly known as *rosin based fluxes*, *organic or water soluble fluxes and solvent removable synthetic fluxes*. These titles are not really logical since rosin based fluxes can be very effectively cleaned off with detergent and hot water, most rosin based fluxes contain organic materials, and organic fluxes can be cleaned off with some solvents. However, these flux family names are known and understood throughout the industry, and for this reason these terms will be used in this discussion of flux types.

It is likely that these classifications will eventually change to indicate the cleaning requirements of the fluxes rather than the materials used in their manufacture. This has already been done in the new IPC document IPC-SF-818, General Requirements for Electronic Soldering Fluxes, where the following flux classifications are used.

L-type fluxes have low or no flux/flux residue activity.
M-type fluxes have moderate flux/flux residue activity.
H-type fluxes have high flux/flux residue activity.

Fluxes will be placed in the appropriate classification according to the results of corrosion and surface resistivity testing. These classifications will be much more useful to the user as they will indicate the cleaning necessary to ensure the ultimate reliability of the product.

Rosin Based Fluxes

As a base, rosin based fluxes have natural rosin, which is obtained from the resin that exudes from certain pine trees. The resin is collected, heated, and distilled, which removes any solid particles, resulting in a purified form of the natural product. It is a homogeneous material with a single melting point. However, it is still the product of nature and varies in composition and quality depending on the tree species from which it was collected and the climatic conditions under which it grew. Because of this variation in the rosin quality there is a move today to develop artificial rosinlike materials that are more consistent in composition.

In spite of the variations in the natural rosin, it has for many years been the favorite flux of the electronics industry. Rosin has the unique property of dissolving copper oxide when it is heated to its molten state. If some pieces of rosin are placed on a sheet of oxidized copper and the metal is slowly heated, the rosin will eventually melt and flow in a thin layer over the surface. This occurs at about 260°F (127°C) and the molten rosin will not burn or char up to 600°F (315°C).

After a while the surface of the copper under the translucent liquid rosin will brighten and, provided the oxide layer is not too thick, will eventually become clean and bright. The active constituent in rosin is abietic acid, which at the soldering temperature is chemically active and attacks the copper oxide, converting it into a copper abiet. As soon as the temperature falls, the rosin resolidifies and reverts back to an inert material, totally noncorrosive, and with a very high electrical resistance. The copper under the flux will stay clean, bright, and shiny no matter how long the metal is allowed to lie around. The solidified rosin now acts as a protective coating and will prevent any reoxidation of the copper surface (Fig. 3.6).

The copper abiet formed by the action of the rosin on the copper oxide is a blue compound and is also inert and harmless. It is sometimes confused with the harmful copper chlorides, which are green. Once the color difference has been seen there should be no difficulty in identifying these compounds.

So with rosin there exists the perfect flux. It removes the oxide from the base copper in the joint, it flows out and allows the solder to move easily, it stays active at soldering temperature, it protects the surface during soldering, and when it cools it is totally noncorrosive and nonconductive. When soldering is complete it will protect the joint from any subsequent contamination. It sounds too perfect to be true; it is. Rosin is a very mild flux. It is the mildest flux in general use and is only effective against very light oxide on copper. It is not able to remove other contaminants and will not even remove heavy copper oxide. In order to solder with rosin the part must be absolutely clean, that is, of excellent solderability.

Pure rosin dissolved in alcohol is called *water white flux*. As mentioned previously, it is totally inert at room temperature and is used where any possibility of corrosion could cause serious problems or where the product cannot be cleaned. It was used very extensively in the manned space program. It still is used in very critical areas of medical electronics and in some military projects. Because of its very mild fluxing action, it is frequently used for solderability testing on the premise that if an item can be soldered with

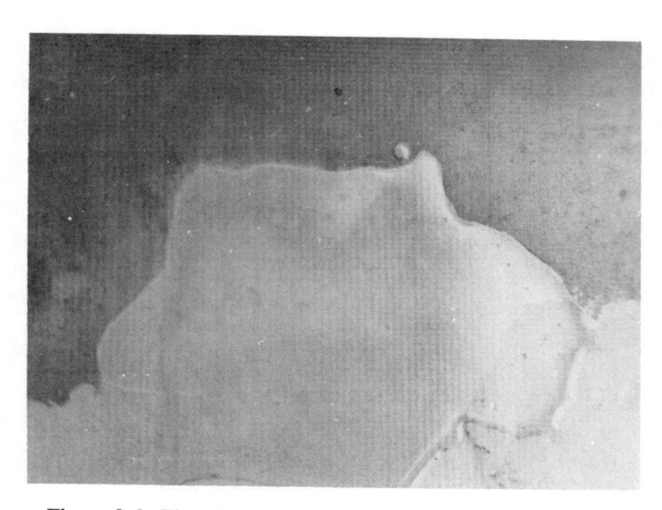

Figure 3.6. The cleaning and protective actions of rosin flux.

this very mild flux, it will certainly be solderable with a more active fluxing agent.

This very mild flux then requires that the materials to be soldered must be extremely clean. In an effort to make this requirement less stringent, a small quantity of a more active ingredient is often added to the flux, and the subsequent mixture of rosin flux and active addition is known as a rosin mildly *active* (RMA) flux. If an even more active ingredient, or a larger amount of the active material is added to the rosin, then the flux is known as a *rosin active* (RA) flux.

These classifications have become generally used in the electronic industry, especially in North America. They were originally developed as identifying names for fluxes in the military specifications of the United States. There are similar classifications in the various European and Japanese specifications, and the appropriate documents must always be used as reference when the product requirements define the flux that must be used.

RMA and RA fluxes must always be cleaned from the assemblies if the highest levels of cleanliness are required. The cleaning system must contain the appropriate solvents to remove both the rosin and the active materials or *activators* as they are usually called. While most of the common cleaning solvents will remove the rosin, for example, trichloroethane or one of the fluorocarbons, they will not necessarily remove the activators, which usually require a polar solvent such as water or alcohol. It is common practice to use a blend of both polar and nonpolar solvents, such as the commercially available azeotropes based on fluorocarbons, or to convert the rosin into a soaplike material with a hot detergent and then wash off with clean water.

When machine soldering was first used it was quite normal to leave on the rosin based fluxes and not to clean the assembly at all. Generally, there was no problem with the performance of the end item. However, the flux was sticky and the product would quickly collect dust and dirt, which were then difficult to clean off. In order to give the product a more pleasing appearance, attempts were made to remove the sticky flux, using the common solvents of those days—trichlorethylene or one of the other degreasing solvents. At this point, electrical and corrosion problems began to appear. These were found to be caused by the exposed activators absorbing moisture from the atmosphere and becoming conductive and corrosive. The rosin had acted as a conformal coating and was most effective in keeping out the moisture, so that the activators, although present on the assemblies, were quite harmless. Cleaning with alcohol solved the problem by washing off the activators. This was a case where no cleaning was much safer than inadequate cleaning.

Organic Fluxes

Organic fluxes, sometimes called intermediate fluxes or water soluble fluxes, are somewhat of a compromise, being more active than any of the rosin based fluxes, yet not as highly corrosive as the acid fluxes used in the metalworking industries. They are generally designed to break down or dissociate at soldering temperatures, leaving only inert and noncorrosive residues. This would be an excellent solution to the fluxing problem if it were only possible to be sure that all the flux actually reached the temperature necessary to be rendered harmless.

In practice this is obviously not possible, and all organic fluxes, no matter what their composition, must be treated as leaving corrosive residues and the assemblies thoroughly cleaned immediately after soldering. There is a wide range of active ingredients used in these fluxes, although the vendor is often reluctant to disclose the exact composition.

The active compounds fall into three groups.

Acids: stearic, glutamic, lactic, citric, etc.

Halogens: hydrochlorides, bromides, hydrazine, etc.

Amides and amines: urea, triethanolamine, etc.

These materials and other parts of the formulation, such as surfactants to assist in reducing the solder surface tension, are dissolved in polyethelene, glycol, organic solvents, water, or usually a mixture of several of these. To repeat, all these fluxes must be considered corrosive, and total cleaning is necessary. There are fluxes coming onto the market that are claimed to be completely noncorrosive, but even these should be viewed with some suspicion until environmental testing, under conditions that the product will see, has confirmed these claims.

In spite of these dangers, organic fluxes have many advantages. They can be cleaned off easily, with no more than hot water, and a well-designed assembly will then have the highest level of cleanliness. Being high activity fluxes they produce excellent soldering results even when the boards and components do not have the highest solderability. The higher activity will also assist in attaining higher soldering speeds because of the reduced solder wetting times.

One word of warning: there is often a belief that by using the higher activity fluxes there is no need to pay the same attention to solderability of the parts, cleanliness in handling, process control, and so on. This is a very dangerous philosophy. Certainly, the use of a more active flux will make the process more forgiving of parts with a lesser solderability, but once the requirement level is reduced, it becomes difficult to maintain. Solderable parts are no more difficult to obtain than those that are not, and bad soldering is an expensive operation. If the decision is made to use a very active flux it should be for many other advantages, not as an excuse to reduce the quality of the soldering process.

During the past years, more especially in 1980–1981, there have been reports of low surface resistivity which is believed to be caused by the overheating of certain of the glycols used as a solvent in fluxes. Excessive heat apparently causes the material to polymerize and form an invisible film on the PWB. This film will not show up in the solvent extract resistivity cleanliness test and is not likely to be detected unless a surface resistivity test or electrical failure brings it to light. The problem has been found with both bare and assembled boards and is believed to be primarily caused by the reflowing of the tin-lead plate, where the higher temperatures cause the fusing fluid to break down. However, there have been reports of this occurring during soldering. The reports are not conclusive, and it is not a common occurrence, but if low surface resistivity becomes a problem it is worth further investigation. This also suggests that excessive time at soldering temperature should be avoided when boards are soldered using fluxes containing these solvents.

Synthetic Activated Fluxes

The advent of the water soluble fluxes developed a demand for a flux of similar activity which could be removed by the use of a suitable solvent such as a fluorocarbon. The Synthetic Activated or SA fluxes resulted, and are now as popular as the water cleanable (organic acid or OA) systems.

When an SA flux is used the comments relating to the OA fluxes also apply. For example, the assemblies must be designed to be cleaned, cleaning must be carried out immediately following soldering, and so on. Any highly active flux must be considered as potentially corrosive, and the soldering and cleaning process properly controlled.

Rosin based fluxes are solid materials dissolved in a solvent; SA fluxes on the other hand are usually based on a totally liquid formula. When the solvent is driven off during preheating therefore it is usually found that the board will be wet or oily and not tacky or dry as with rosin based systems.

Low Solids or "No Clean" Fluxes

Cleaning the flux residues from the printed circuit assemblies is a costly process no matter which system is used. Solvents are extremely expensive both to procure and to dispose of when they become contaminated. Clean water is not cheap, and again the problem of effluent disposal becomes more and more difficult. The development of the "Low Solids" or "No Clean" fluxes offers the possibility of eliminating all of these costs completely. If used correctly the manufacturers claim that the assembly coming from the solder machine will have the appearance of a newly cleaned board with no visible or sticky residues. In addition when tested for cleanliness it will comply with most of the current cleanliness standards.

These fluxes are similar in activity to an RMA flux and have a solids content of about 15%. The solderability of both boards and components must be excellent and they must both be scrupulously clean before soldering. This entails procuring clean parts and providing excellent housekeeping throughout the storage and assembly of the product. Low solids fluxes offer considerable cost potential. However, as with any change to another flux type adequate testing should be carried out to ensure that there is no potential jeopardy to the long-term reliability of the product.

Fluxes in General

No matter which type of flux is chosen, the listing of any flux vendor will show many alternatives. These generally are related to the solvent used, the viscosity, and the ratio of flux to solvent, more often referred to as the *solids content*. There will be special formulations for the foam fluxer, containing a foaming agent. Flux vendors are happy to assist in the correct selection, and their expertise and experience should be used to the fullest. There is no way that this book can list all the available fluxes or do more than give some basic guidelines for flux selection. Much will depend on the design of the board and components as discussed earlier.

However, there are some rules of thumb that can be used as a basis for flux selection. Use the lowest solids content flux that will provide satisfactory soldering, at the throughput speeds required. If there is a problem with icicles or bridging, use a higher solids content. If there is a difficulty in obtaining a fast enough soldering speed try a more active flux. Remember that the higher the solids content and the more active the flux, the more difficult will be the cleaning after soldering. Fluxes with an alcohol solvent will require more frequent density checks and thinner additions than fluxes with less volatile solvents.

In summary, the following actions must be taken when selecting fluxes:

Consider the design of the board and components and the type of cleaning that can be used.

Discuss the requirements with several flux vendors; ask for samples and try them out in production.

Once the selection has been made, do not change fluxes without repeating the tests.

Use only the vendor's recommended thinners.

Make the tests on the board with the closest conductor spacing.

For the tests use the actual soldering machine that will be used in production.

Polar and Nonpolar Contaminants

At this point, having discussed the relationship between flux selection and cleaning of the soldered assembly, it is time to step aside from the soldering process and explain the terms polar and *nonpolar* as they relate to flux residues and other contaminants that have to be removed from the PWB.

Polar or ionic contaminants can be defined as compounds that, in the presence of water, will be conductive and can corrode metals. Some typical polar contaminants are salts from plating, salt from sweaty fingers, and flux activators such as chlorides. These contaminants can only be removed by using polar solvents, which include alcohol, ethanol, glycol-ether, and water.

Nonpolar contaminants are not conductive, are not soluble in water, and include oils and greases, which can be picked up from handling or prior processes, and rosin from the flux residues. These contaminants can cause problems with electrical contacts, will retain dust and other dirt, and prevent the adhesion of solder mask and screened-on legend. They will not be removed by most polar solvents; for example, water will not wash off grease. Nonpolar solvents have to be used, such as perchloroethylene, 1,1,1-trichloroethane, or other chlorinated or fluorinated solvents.

Unfortunately, it is not likely that any assembly will have only one form of contamination after soldering. If an RMA or RA flux is used, there will be nonpolar rosin and polar activators. Even if a water white rosin flux is used, unless great care is taken in handling, ionic contaminants will be picked up from fingers, the atmosphere, contacting benches, containers, and tools.

Cleaning, therefore, requires the use of materials that will remove both forms of contamination. These include proprietary solvents especially formulated to contain both polar and nonpolar solvents or a dual wash system, for example, perchloroethylene to remove nonpolar soils, followed by an alcohol wash to remove the ionic contamination. A popular alternative uses a hot detergent solution to break down the greases, oils, and the rosin that form the nonpolar contaminants, followed by a plain hot water wash to remove the detergent, the residues, and the ionic contamination.

As already discussed, the method of cleaning and the solvents used have to take into account the design of the PW assembly and the components.

Chapter 4

Solders

In the machine soldering of electronic assemblies the solder performs two functions: it heats the metals in the joint to bring them to soldering temperature, and then it provides the mechanical bond to lock them together with excellent electrical conductivity. While any molten metal will perform the first function there are some very particular requirements necessary for the solder to form a reliable mechanical and electrical connection.

THE TIN-LEAD ALLOY

Most soldering in the electronics industry is carried out with a tin-lead alloy. Rarely, and only for some very special purpose, is another alloy used, and this will be reviewed briefly later in this chapter. The alloying of tin and lead produces a metal with some properties that are quite different from those of either metal alone.

The so-called *phase diagram* of Fig. 4.1 shows the behavior of the mixture of tin and lead as it is heated, for different ratios of the two metals. Pure lead melts at 621°F (327°C), while the pure tin melts at 450°F (232°C). The mixture of tin and lead will melt at 370°F (188°C), when the ratio of tin to lead is approximately 60/40. It can be seen from the phase diagram that for intermediate ratios of tin-lead there is a range of melting temperatures, and that as the ratio moves from the 60/40 area, the solder goes through a plastic or pasty stage instead of changing directly from solid to liquid.

The tin-lead ratio at which the solder alloy goes directly from solid to liquid is known as the *eutectic* composition, and that alloy is known as the eutectic alloy. As well as not having a plastic range, the eutectic alloy has the lowest melting point of all the tin-lead alloys. This alloy is generally referred to as 63/37, with a 63% tin content. In fact it is extremely difficult to determine exactly the tin-lead ratio for the eutectic alloy. Minute impurities and other factors come into the equation, and there is good evidence to

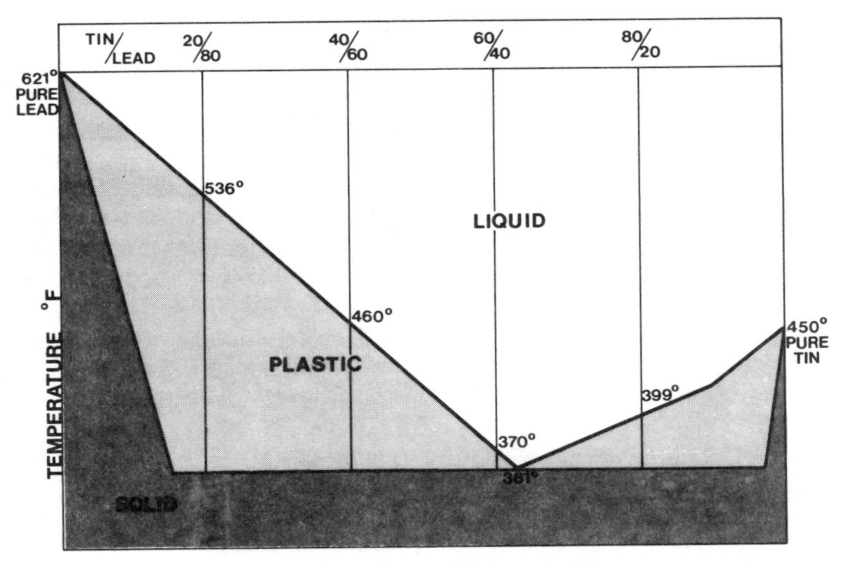

Figure 4.1. Simplified tin–lead phase diagram.

suggest that the correct alloy is nearer to the 60/40 ratio. Fortunately, this is not of real concern to the machine soldering process, as there is little or no discernible difference between the 60/40 and the 63/37 alloys. In this book, therefore, when eutectic solder is mentioned it will refer to the generally accepted alloys in this range.

The eutectic alloy is preferred for machine soldering for the following reasons:

Soldering can be carried out at the lowest possible temperature.

Eutectic solder wets solderable surfaces easily and quickly.

The eutectic alloy produces smooth shiny solder joints that are easy to inspect, and any changes to this appearance indicate incipient soldering problems.

This solder is mobile and flows readily.

Because it changes directly from liquid to solid there is less possibility of the joints being disturbed during cooling.

SOLDER ALLOY SELECTION

Although the eutectic alloy has many advantages, there is always a temptation to attempt to use a lower tin content solder to reduce costs, since the price of solder is directly related to the tin content. This can be done, but only if the boards and components have excellent solderability, the process is well controlled, and the board conductor spacing is not extremely small. Any change from the eutectic alloy must be carried out with extreme care and careful records maintained of all the tests that must be carried out before such a change is implemented. As will be seen in Chapter 7, any increase in defective soldered joints is expensive, and the problems may only be discovered sometime after the soldering has been done. Even with the use of eutectic solder the cost of machine soldering is extremely low, while the cost of joint inspection and touch up is extremely high. An increase of only a tiny part of 1% will more than offset any savings that will accrue from the use of a lower cost alloy.

It can be done, but move very cautiously, keep meticulous records, and consider the entire cost of the operation when making any cost analysis. The following results must be expected.

A reduction in soldering speed
An increased soldering temperature
A less shiny joint
An increase in the number of defective joints

Another consideration in the choice of solder alloy is the purity of the solder as purchased. There is a certain small cost advantage in buying the less pure alloy, but it will become apparent that this saving is probably not real when the total cost of the process is considered.

Unfortunately, this is a much more complex problem than the choice of the solder tin-lead ratio. Very small levels of some contaminants in the solder bath can cause major effects on the soldering process. These levels are extremely difficult to measure. Some contaminants can offset the effects of other impurities, so that combinations of contaminants may have very different effects from those seen when any one of them is present alone. This is explained in more detail in the next section.

It is much safer and much easier to control the process if the solder bath is initially filled with solder of a known high purity. Starting in this way will enable the control of the pot to be set up and monitored with the certainty that no impurity was introduced during the initial filling. Once operating satisfactorily any reduction in the purity level can be permitted in a controlled manner. Any processing problems or increasing solder joint failures can then be controlled and the necessary corrective actions taken. Process changes will be engineered, not left to chance.

CONTAMINATION LEVELS AND CONTROLS

There are many specifications controlling the levels of impurities in the "as purchased" solder.

From the United Kingdom: BS 219 and BS 441, soft solders
From Germany: DIN 1707, rosin cored solder wire
From the United States: ASTM 571 and QQS 571 E, solder metal
From Japan: JICS 2512, rosin cored solder wire

However, the limits of the contaminants that can be accepted in the solder pot before the joint will be affected are not precisely defined, and there is little agreement on the figures that should be used. This is not to say that any of the specifications of contaminant levels is incorrect, but this demonstrates the complex nature of the situation. The testing involved in checking these contaminants in the laboratory is difficult, and the contamination is so small that measuring the level approaches the limit of the available technology in many laboratories. When an attempt is made to relate the contaminant level to actual solder joint failure, the task of inspecting the joints becomes almost impossible. The concern is to eliminate defective solder joints, which means that even a failure rate of 0.05% is not acceptable. This requires inspecting 10,000 joints to find only 5 that are defective. Human judgment and other subjective factors also enter into the inspection criteria, so that any results must be suspect.

There have been many attempts to scientifically determine the acceptable metallic contamination levels in solder which will produce smooth, bright, shiny joints. These attempts have invariably failed to provide any clearly defined figures, because of the many variables in the process and the apparently impossible task of relating the contamination levels to the appearance of the solder surface.

Probably the most work on the subject has been carried out by the International Tin Research Institute and published in the paper "Effects of Certain Impurity Elements on the Wetting Properties of 60% Tin 40% Lead Solders," by N. L. Ackroyd, C. A. Mackay, and C. J. Thwaites. This excellent paper covers all the main contaminants that are found in practice in the pot of a solder machine. However, this work was all carried out in the laboratory, on small samples, and under carefully controlled conditions. There have been challenges made to some of the data on the grounds that in practice much higher levels of contamination can be accepted than those proposed by the authors without any deleterious effects on the soldered joints.

There are so many variables in the soldering process and in the materials to be soldered that it is doubtful if this controversy will ever by resolved. A contaminant level that may be perfectly acceptable for a solder masked board with widely spaced component mounting pads may be totally disastrous for a board with many exposed parallel conductors, closely spaced, where the mobility of the solder is extremely important if shorts between conductors are to be avoided. Therefore, it is vitally important that the process be set up with clean solder of a known purity and composition, and

Table 4.1 Solder Contamination Levels[a]

Contaminants	IPC 815A Mil Spec[b]	User A	User B	User C	Proposed Starting Point
Aluminum	0.006	0.0005	0.006	0.001	0.005
Antimony	0.2–0.5[c]	1.0	—	0.1	0.50
Arsenic	0.03	0.20	0.03	0.03	—
Bismuth	0.25	0.50	0.25	0.25	0.25
Cadmium	0.005	0.15	0.005	0.005	0.005
Copper	0.30	0.29	0.25	0.25	0.20
Gold	0.20	—	0.05	0.005	0.10
Phosphorus	—	0.10	—	0.01	—
Sulfur	—	0.0015	—	0.002	—
Silver	0.10	—	0.10	0.2	0.10
Nickel	0.10	—	—	0.01	0.10
Iron	0.20	—	0.02	0.01	0.20
Zinc	0.005	0.003	0.005	0.001	0.003
Others	[d]	—	—	—	0.05

[a] All figures are in percentages by weight.
[b] For all specifications tin content to be between 59.5 and 63.5%.
[c] With the exception of antimony in the first column, all figures are maximum levels.
[d] Total of gold, cadmium, zinc, and aluminum not to exceed 0.40%.
Total of gold, cadmium, zinc, aluminum, and copper not to exceed 0.40%.

that these factors are controlled so that variations in them can be measured and recorded. Effects on the soldering results can then be related to the changes in the solder and any necessary corrective actions taken (Table 4.1).

It is necessary to understand the effects on the soldering process and to know the visual appearance of joint failures caused by solder contamination. Quite small additions of contaminants can affect the wetting of the joint, the speed of wetting, and the ability of the solder to wet the metal completely. The mechanical strength of the joint can also be affected through changes to the crystalline structure of the solder.

Although there are many possible contaminating materials, it is only necessary to consider those that are likely to be found in the electronic assembly operation or that can be introduced by the components to be soldered. Each soldering installation should be reviewed to be sure that some material other than those discussed cannot find its way into the solder pot. As already mentioned, there is no general agreement on the acceptable levels of contamination, and those given in this book represent the most conservative approach to contamination control. At the levels quoted the visual appearance of the joint will begin to change, without producing unreliable connections. As will be seen, these levels may possibly be changed later depending on particular circumstances, and the soldering results obtained.

Copper

Mass soldering of electronics involves copper in the component leads and the PWBs, and therefore it is one of the most common contaminants found. As the level of copper increases, the solder joints begin to exhibit a gritty appearance, as if there were fine sand under the solder surface. There is good evidence to suggest that the wetting ability of the solder is slightly reduced. Some authorities set the maximum level at 0.3%; others, as high as 0.4%. The safe level is lower than either of these, and a maximum of 0.2% is the proposed specified figure. Anytime that the joints are seen to have a gritty appearance, copper contamination should be the prime suspect.

Aluminum

Aluminum can be picked up from pallets, fixtures, and even from the abrasion of worn conveyor rails. Very small levels of aluminum can cause gritty looking joints and an increase in the rate of oxidation of the solder surface, which can show up as an increase in the amount of dross on the pot. In addition, the solder will not have such a bright shiny surface.

All aluminum that comes into contact with the solder must have some protection to avoid contaminating the pot. Anodizing, a PTFE (polytetrafluoroethylene; Teflon® is the trade name used by DuPont) coating, or a good epoxy paint are some of the protective systems used.

It has been found that a small addition of antimony will eliminate the effects of aluminum contamination by combining with the metal to form a dross that remains on the surface of the solder. Since most solders contain a small percentage of antimony this may explain why aluminum contamination is rarely found to be a problem, and when it does occur it is usually because of some major accident that has dumped a considerable amount of the metal into the solder pot. In spite of this, however, it is unwise to permit anything containing unprotected aluminum to contact the solder. The maximum specified level is 0.005%.

Gold

Gold was at one time the major contaminant to be found in many solder pots. It was one of the preferred finishes on pins, connectors, and even PWBs. The eventual discovery that the tin–gold–copper intermetallic could promote the onset of solder cracking in the joint caused the gradual demise of gold as a solderable finish, and a dramatic increase in the price of gold finally killed it off. Gold is still sometimes found on component leads, and if many are soldered it is wise to include this metal in the possible sources of solder contamination. It will make the joints have a frosty appearance, and if the level is high the joints will also look gritty. Gold contamination can also reduce the wetting ability of the solder. A maximum contamination level of 0.1% is specified. There is no need to retain solder contaminated with gold, the solder vendor will always purchase scrap solder based on an assay, and gold contamination will frequently bring a price equal or greater than the cost of new solder.

Cadmium

Cadmium is usually introduced into the solder pot from cadmium plated hardware used to mount large or heavy components. The hardware must be passivated, coated with a solder stop, or exchanged for stainless steel to prevent this contamination. Cadmium is a particularly bad material to get in the solder. It causes a reduction in the wetting power of the solder and produces a very dull looking joint. Very small amounts of cadmium will cause this effect and the maximum allowable level is set at 0.005%.

Zinc

Zinc is another metal that can be introduced from the finish on hardware, and the same remarks as for cadmium apply. Another source of zinc is from the soldering of brass parts: terminals, pins, fastenings, and so on. The tin in the solder is capable of leaching out the zinc from the brass. Zinc in very small levels will cause the solder joint to have a rough frosty appearance and will also cause excessive dross. The specified maximum level is 0.003%.

Antimony

The presence of antimony has been shown to reduce slightly the wetting power of solder and to increase slightly the wetting time. Up to 0.5% these effects are scarcely noticeable. Some specifications require a minimum amount of antimony in the purchased solder (for SQ-S-571 it is 0.2%) to improve the low temperature characteristics of the soldered joint. Without antimony, it has been suggested that low temperature aging can cause the solder to crumble. This is primarily of concern in military and aerospace applications. Too much antimony, on the other hand, can cause brittle joints; the solder becomes sluggish and can contribute to solder shorts, icicles, and excessive solder on the joints. The maximum level should not exceed 0.5%.

Iron

At temperatures above 806°F (430°C) the solder will dissolve iron from the surface of a new solder pot that is not given some form of protective coating. Once in use for a short time an oxide surface will form naturally on the walls and prevent further dissolution. This intitial contamination is more prevalent with cast iron pots and can cause excessive dross. This will quickly clear of its own accord without any other action except to keep the surface clear of dross. It is rare to find iron contamination in any other circumstances, unless very high solder temperatures are used, above 850°F (454°C). However, it is wise to be sure that any tool used in the solder, such as scrapers or dross removable tools, are fabricated from stainless steel. The maximum limit for iron is 0.02%.

Silver

Silver is another metal that finds its way into the pot from the plated leads of components. As with gold the use has been reduced following the price increase of the metal. Silver is one of the least detrimental of the contaminants and is in fact deliberately added to some solders. This will be discussed later. It can cause a dull appearance to the joint and can, in high concentrations, reduce the mobility of the solder. The specified maximum level is 0.1%.

Nickel

Nickel is another contaminant that can be picked up from component leads or plated hardware. It is not commonly found in the solder pot, but it will cause small blisters on the surface of the joint and form insoluble compounds in the solder bath. The maximum limit for nickel is 0.01%.

Other Contaminants

Many other materials can contaminate the solder. Some will cause soldering problems and reduce the reliability of the joint. In a well-kept shop they are unlikely to be found in the solder machine and are therefore lumped together as "other contaminants." They include arsenic, bismuth, indium, phosphorus, and sulfur. Not all have a bad effect on the joint; indeed, some are deliberately added to the solder alloy for special soldering work. For example, indium, cadmium, zinc, and bismuth are all alloyed with solder to produce special features that are required for particular joining purposes. However, here the soldering of PWBs by machine is under consideration, and these alloys are rarely, if ever, used.

As an additional complication to this already complex problem of solder contamination, it must be realized that when two or more contaminants are present the results may not be the same as those obtained from either contaminant by itself. Little work has been done on this effect although the combination of aluminum and antimony was mentioned earlier in this chapter.

It is absolutely necessary to have a good understanding of the problems that can arise from contaminants in the solder, but it is even more important to take the necessary precautions to prevent this problem in the first place and to have good control of the process so that if it does occur the solution can be found quickly and confidently.

TIN–LEAD RATIO

There is evidence from practical experience on the shop floor that the tin–lead ratio is much more important than previously believed when the objective is to produce bright, shiny soldered joints. The findings suggest that small amounts of metallic contamination in the solder shift the eutectic point and thus introduce a small "pasty phase" in what would otherwise be considered a eutectic solder. As the alloy cools through this phase excess lead tends to produce a crystaline formation on the surface thus causing a frosty or, in extreme cases, a gritty joint (Fig. 4.2).

For example, it is not uncommon to see frosty joints when the copper contamination level is high and the tin ratio is low (Fig. 4.3), even though both are within the normally specified limits. The addition of tin will almost

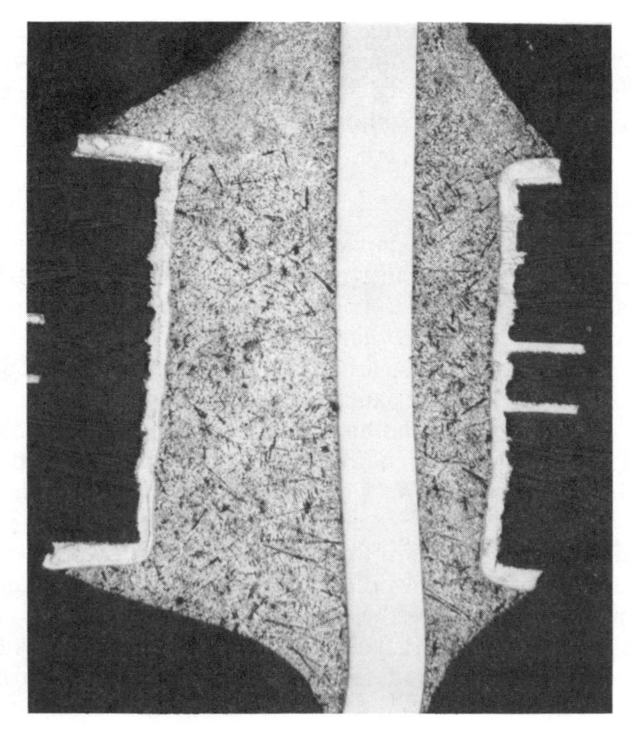

Figure 4.2. Cross section of a solder joint showing needlelike lead dendrites which caused a gritty appearance on the solder surface. This was caused by contaminated solder.

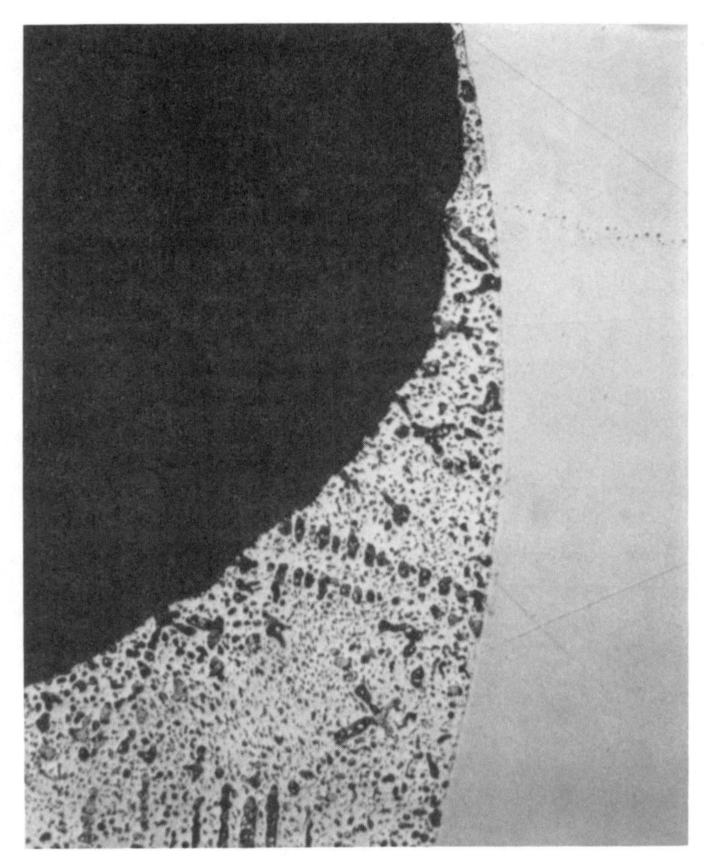

Figure 4.3. Enlarged cross section of the solder fillet showing lead dendrites reaching the solder surface and causing a gritty appearance.

always eliminate the frosty appearance. Of course there is little or no evidence to show that frosty joints are any less reliable than those that are bright and shiny (Fig. 4.4).

SOLDER CONTAMINATION CONTROL

The first line of defense in controlling contamination is to maintain excellent housekeeping of the solder machine, fixtures, pallets, and anything else that can introduce contaminants into the solder pot. However, it is impossible to avoid the metals of the PWB and the component leads from passing through the solder, and this is where most of the unwanted metals will come from. Every attempt must be made to keep the assemblies clean and free from oil,

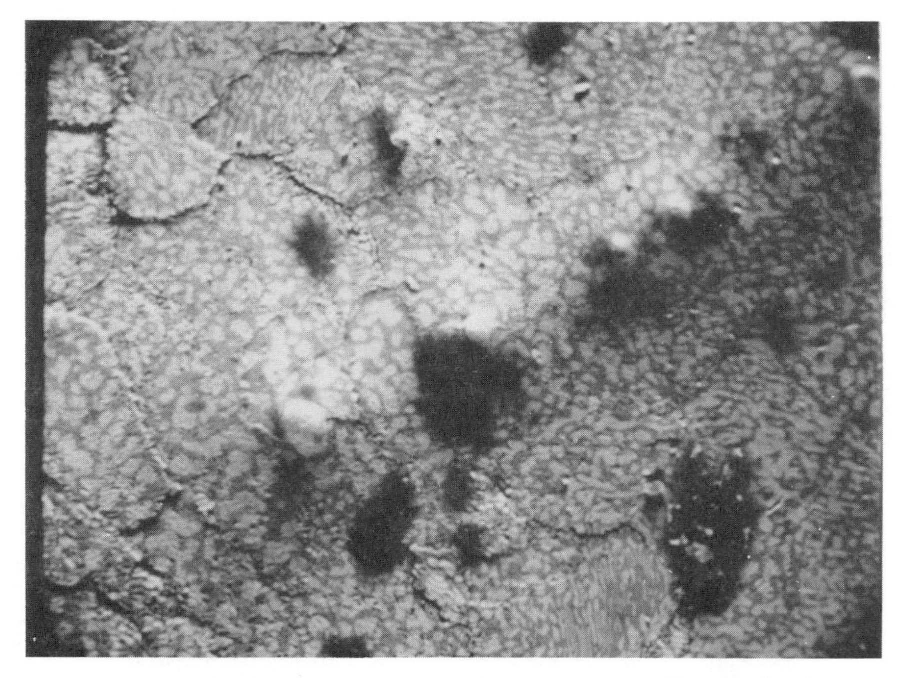

Figure 4.4. Scanning electron beam microscope view of the solder surface showing the emerging lead dendrites.

grease, and other nonmetallics. These will not be likely to affect the solder but there is the possibility, and they will certainly contaminate the flux.

Even with all these precautions it is absolutely necessary to have the solder analyzed on a regular basis. Careful recording of the results of these analyses will provide invaluable help in troubleshooting soldering problems and will clearly indicate the onset of an out of tolerance situation before it can cause joint failure.

When the soldering operation is started with a clean pot of pure eutectic solder, as each assembly is soldered it will add a small amount of contamination. Some copper will be dissolved from the PWB, perhaps a small amount of gold from a transistor lead or some nickel from a resistor; the possibilities are numerous. At the same time the board will take solder from the pot to form the actual joints. So the amount of solder in the pot will be reduced while the contamination level will increase. After some time fresh solder will be added to the pot to bring the solder level back to its original point. This will reduce the percentage of contaminants. In addition, as the amount of contamination increases, it will reduce the ability of the solder to dissolve further contaminating metals, and eventually the solder will reach a steady state with the amount of contamination that is being added being balanced by the fresh solder that is used to replenish the bath (Fig. 4.5).

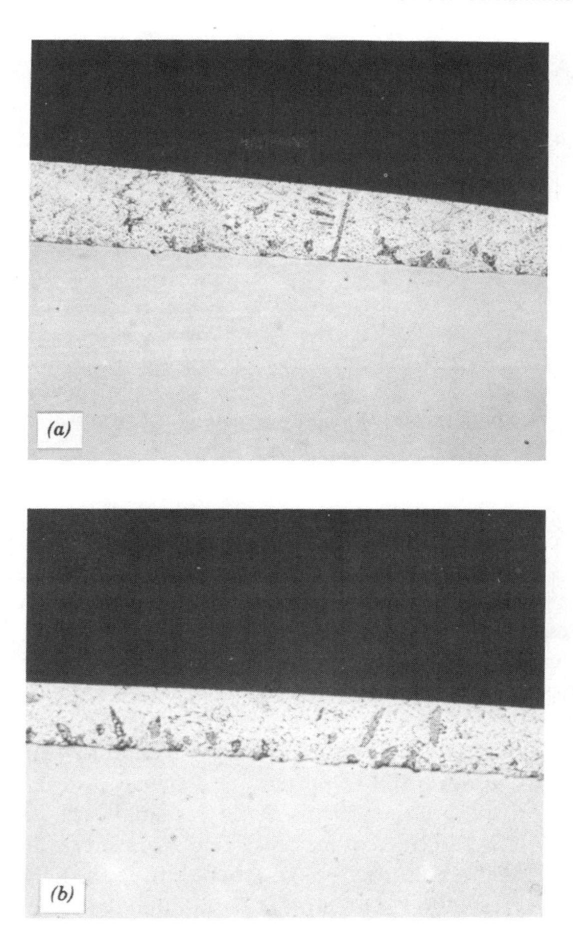

Figure 4.5. (*a*) Cross section of a solder coating on a DIP lead showing the needlelike lead dendrites due to solder contamination. (*b*) A similar lead coated with clean solder.

When all these factors are combined, it can be seen that the contamination level will gradually increase with the curve slowly leveling off (Fig. 4.6). The exact shape and the point at which it finally flattens is dependent on many factors, including pot size, board and lead materials, and housekeeping. As these factors change the contamination level curve will also change. There is also the very real danger of introducing contaminants by accident or carelessness. All these items can cause changes in contamination levels, often unexpectedly.

Starting with clean solder the analysis should be carried out every week, of every 10,000 ft^2 of board soldered, whichever is the longer period. The results of the analysis must be plotted as shown in Fig. 4.7. Most solder vendors offer an analysis service if this cannot be done locally. First, each

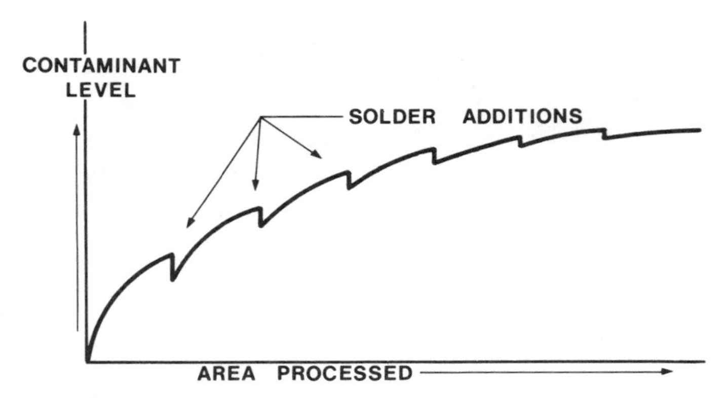

Figure 4.6. A graph showing the growth of solder contamination.

individual element given in the analysis should be plotted, together with the tin content. Very quickly, a pattern will be established, and then only those elements of a significant quantity need be monitored individually. Next, the remainder can be lumped together under Other Contaminants, and the total figure reviewed.

Once the curves have flattened, the analyses need not be carried out so frequently. First, double the time between checks, then extend the intervals still further. Very much now depends on the particular conditions existing in the operation. If the level of contamination is well below the desired figures then there is a good margin of tolerance, and this allows the testing interval to be increased. If the volume of production is small, it is better to base the testing interval on the number of assemblies processed rather than a particular time interval. If the volume of product is high, then the analyses must be carried out more frequently, and a regular schedule based on time is usually easier to arrange.

The jeopardy always exists that the incidence of faulty joints can increase without becoming noticeable until it reaches such a volume that the only solution is to shut down the operation until the cure can be found. This is costly in lost production, unreliable products, and unnecessary rework. Compared with this the cost of regular solder analysis is well worth the price. Under normal conditions it should be carried out at least once each month or every 50,000 ft^2 of boards soldered. If there is a change to the product, for example, going from solder coated boards to bare copper boards, then the analysis should be performed more frequently until the contamination pattern is once more established. Whenever dull, gritty joints are seen, or the incidence of shorts or icicles increases, or if dewetting or excessive solder causes solder joint rejection, carry out an immediate solder analysis. If there is any doubt as to the state of the solder and an analysis cannot be obtained quickly, it is better to dump the pot and refill with solder of a known purity.

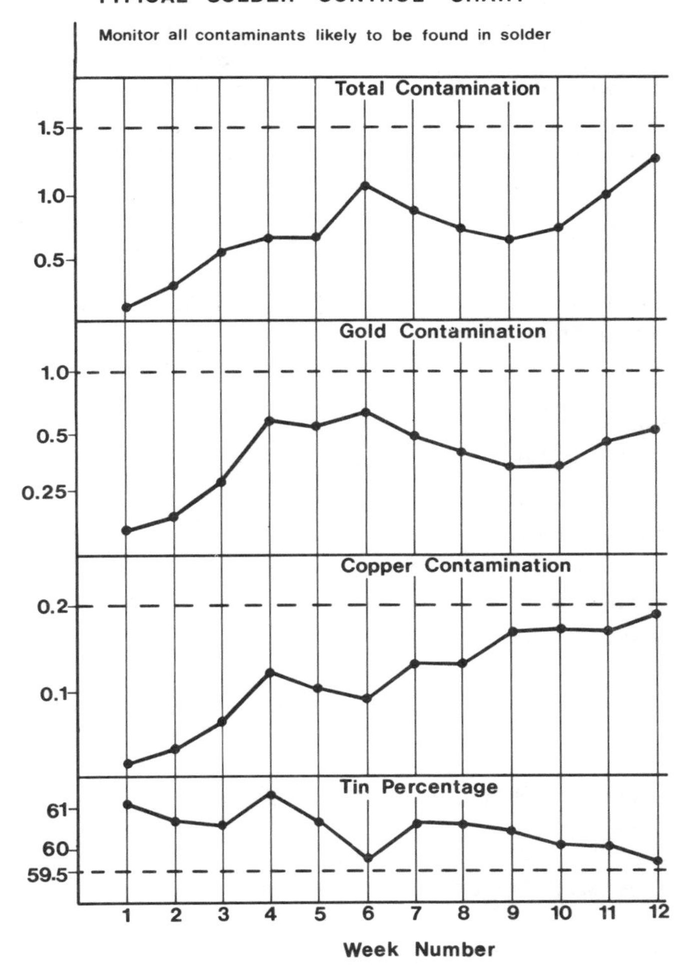

Figure 4.7. A typical graph of solder contamination levels.

While the frequency of solder analysis is not a subject of great argument, the question of actual acceptable levels of contamination certainly is. Table 4.1 has a column headed Proposed Starting Point. The figures here are the same as those given in the descriptions of the problems that can be expected from the various contaminating metals. The figures are conservative, and many experts will claim that they can be exceeded without causing any soldering problems. Depending on the product and the board design this may well be true.

Dumping and refilling the solder pot can be expensive, even though the solder vendor will always buy the rejected solder. The maximum level of solder contamination must not be set higher than necessary in order to avoid this additional expense. If the experts cannot agree on the correct limits, how is the local engineer to make this determination? The answer lies in the accurate records of the solder analysis which can then be compared with the actual soldering results.

Even if the contamination curve eventually levels out above the suggested maximum for any particular contaminant, if the soldering results are not affected there is no reason why a higher limit should not be set. The figures given are a starting point and can be adjusted as experience is gained. They are conservative so that no matter what the type of board to be processed, solder with less than this level of contamination can have no deleterious effects on the soldered joint.

The following are the key points in the selection and control of the solder:

Use a eutectic tin-lead alloy.

Purchase the purest alloy possible.

Make any alloy changes carefully with tight process control.

Monitor contamination levels regularly.

Maintain accurate records of contamination and alloy composition.

Make an immediate solder analysis when soldering problems arise.

If in doubt dump the solder and refill the pot with fresh solder of a known purity.

Even if the solder analysis shows no change continue regular analyses.

OTHER SOLDER ALLOYS

It is rare to find any other alloy than tin-lead used for mass soldering in the electronics industry. There are some other alloys used for specific purposes, which will be discussed briefly. If the need arises for an alternative alloy it is wise to discuss the requirement with a solder vendor who will be able to provide detailed information. The possibilities are too broad to be covered here.

When it is necessary to solder connections to ceramic components with deposited silver conductors, the tin-lead alloy will dissolve the silver leaving the substrate bare. If a small percentage of silver, 2–3%, is added to the solder, it will reduce the ability of the alloy to dissolve the silver on the substrate and allow the connections to be made without difficulty. The resulting joint will not have the shiny appearance of the eutectic solder but will be perfectly sound.

Indium solders are used where special mechanical characteristics are required of the joint. For example, lead-silver-indium will provide a high temperature joint. An indium-tin alloy will adhere to glass. Indium-lead has a much lower solubility for gold than tin-lead and can be of use in soldering gold plated connections.

Bismuth alloys, on the other hand, offer the advantages of extremely low melting points—as low as 158°F (70°C) going up to 440°F (227°C). These alloys can be useful when the components to be soldered would be damaged by higher temperatures.

Cadmium-zinc solders are used to solder aluminum and produce joints of reasonable strength when used with the correct flux. All these soldering systems are used for special purposes only, and it is extremely unlikely that they would ever be used in machine soldering. However, it is useful to have an understanding of the possible uses of other alloys. One of them may at some time solve a particular problem.

It is much more common to find other alloys of tin-lead used in the electronics industry. For example, it is sometimes necessary to assemble and solder in two sequences. Some components may be assembled and soldered to the board, and then another operation is carried out. Finally, the remainder of the components are assembled, and it is necessary to resolder the board without disturbing the original joints.

It is possible to make the first soldering with a low tin content solder, say 20/40, with a liquidus temperature of 550°F (288°C) and then carry out the final soldering with a eutectic alloy with a liquidus temperature of 361°F (183°C). Note that the 20/40 alloy becomes pasty at 361°F (183°C), so that this procedure is not totally reliable and requires very close control of the soldering temperature.

Chapter 5

The Solder Machine

There are on the market many different types of soldering machines. It is impossible in this book to describe all of them, but fortunately they all have certain features in common, and these will be reviewed as listed below. At the end of this chapter features that are particular to an individual make of machine will be discussed.

The common features are as follows (Fig. 5.1):

The various equipments for applying flux to the assemblies (*fluxers*)
Methods of heating the board prior to soldering, and the devices used (*preheaters*)
Methods of heating and containing the solder (*solder pot*)
Systems for forming the wave (*pumps and nozzles*)
Devices for moving the assembly through the machine (*conveyors*)
Methods of controlling the system (*controls*)

FLUXERS

As discussed in previous chapters, flux is an essential part of the soldering process, and the application of flux in a controlled manner is an important part of the soldering process. Flux can be applied to the assembly by painting it onto the underside of the board with a brush. This is a useful method if only a very few boards are to be soldered, or when testing different fluxes, and production cannot be stopped to wash out and refill the fluxer, or only a small sample quantity is available. It is, of course, not a very controllable method, and is totally inadequate for quantity production.

Several methods of automatically applying flux have been developed; each has its own advantages and disadvantages, and each will be described at length in this section. Of all the methods however, foam fluxing is without

Figure 5.1. A solder machine under construction showing the various modules supported on the conveyor frame. Courtesy Electrovert Ltd.

doubt the simplest and preferred, unless some particular factors prevent its use.

The Foam Fluxer

The foam fluxer (Fig. 5.2) is simple to operate, the setup is minimal, and with regular, but nominal, maintenance it will produce excellent results. The foam fluxer, as its name suggests, has to be used with a foaming flux. This is simply a flux that will produce a stable head of foam, that is, a mass of tiny bubbles when air is forced through the liquid, either because of the normal property of the ingredients or through the addition of a detergent or other "foaming agent." Any chemical added to assist in foaming must not interfere with the fluxing action or cause problems in cleaning the assembly after soldering. There are many fluxes of all types especially formulated for use in the foam fluxer, so that there is no difficulty in obtaining the correct selection for any requirement.

The foam fluxer is an extremely simple device and operates in the same way that a child can overflow a glass of milk by blowing into the fluid with a straw (Fig. 5.3). Inside the foam fluxer nozzle there is a porous tube, with one or both ends connected to a source of compressed air, and the interior of the tube therefore under pressure. The air is forced out of the tiny pores of the tube and produces bubbles in the flux. These bubbles rise up into the

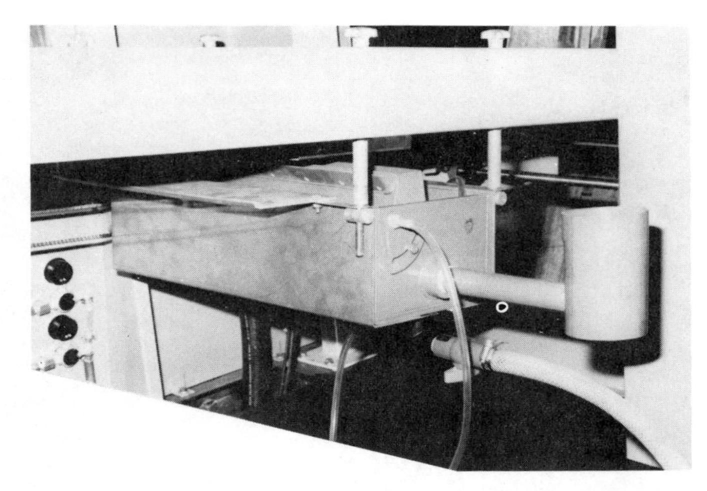

Figure 5.2. A foam fluxer hung on the solder machine conveyor. Courtesy Electrovert Ltd.

fluxer nozzle and are forced farther and farther up by the pressure of the continuing stream of bubbles. Thus, a mass of foam is formed which slowly overflows the top of the nozzle and produces a standing head which protrudes about 1/4 in. (6 mm) above the nozzle edge. The foam eventually overflows and runs down the outside of the nozzle back into the fluxer. The board is conveyed over the head of foam so that the underside of the assembly just passes through the mass of flux bubbles and is coated with a thin but consistent layer of flux. If the height of the assembly over the foam head is correctly adjusted there will be no flux deposited on the top or component side of the board.

To avoid putting an excessive amount of flux on the board, an air knife or a brush is usually fitted on the exit side of the nozzle. The air knife requires a supply of compressed air but is generally preferred to the brush (Fig. 5.4). It must be adjusted so that the air stream blows back the excess flux, which then drops back into the fluxer. It must not affect the foam head and is usually directed about five degrees from the vertical pointing away from the fluxer. Unless the brush has quite stiff bristles, it quickly becomes soaked with flux and requires frequent washing out with thinners to restore its effectiveness. The brush should be adjusted so that it touches only the tips of the protruding leads. It should not brush along the surface of the board. Stiff bristles, on the other hand, can cause movement of the components if they are not clinched or otherwise retained in the board. It is wise to remove all surplus flux. First, it reduces flux usage. Second, it helps to maintain the cleanliness of the machine, but also it prevents the possibility of fire from flammable flux dripping onto the preheaters.

Other than the mildest of rosin based fluxes, all contain chemicals that can attack the component parts of the fluxer. They are therefore usually made of

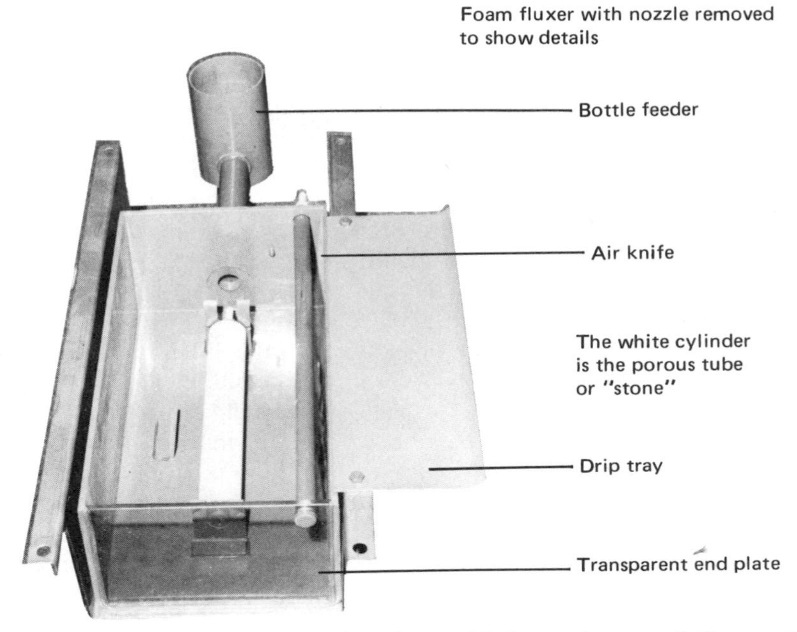

Figure 5.3. An interior view of a typical foam fluxer with the nozzle removed. Courtesy Electrovert Ltd.

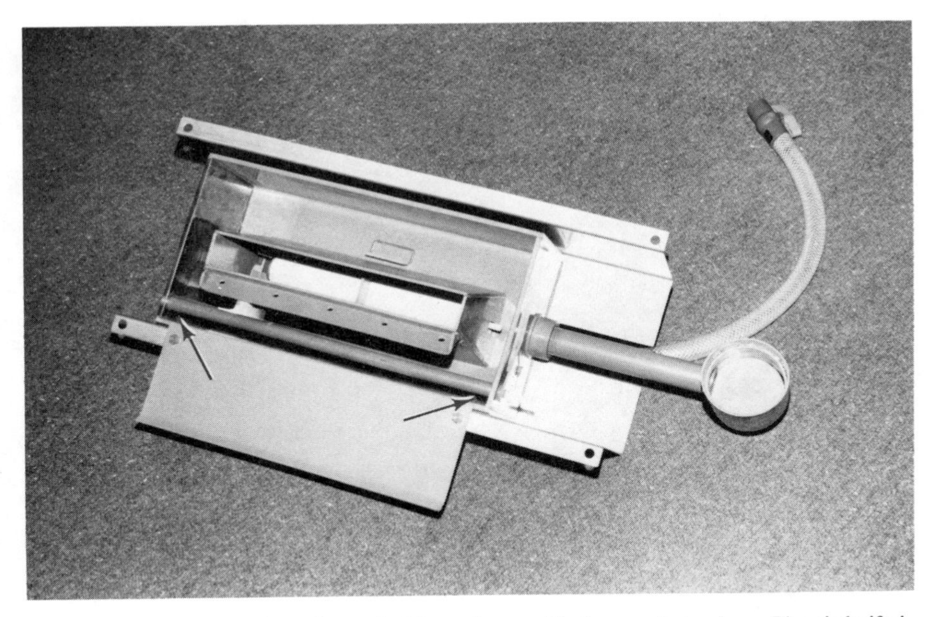

Figure 5.4. An interior view of a typical foam fluxer with the nozzle in place. The air knife is identified by the arrows.

stainless steel, titanium, or one of the plastics such as polypropylene. Stainless steel and titanium are extremely durable and have a very long life, but they are more expensive and require expert welding and fabrication. The greater use of the organic fluxes, which are extremely active and can attack some of the grades of stainless steel, has also helped to produce a swing over to plastic for fabricating fluxers. Polypropylene and polyvinyl chloride (PVC) are two of the materials most frequently used, and a transparent version of the latter makes a most attractive fluxer with the additional advantage that the flux level is visible at all times. However, fluxers of these materials must be protected from the heat of the preheaters or they will warp or, in extreme cases, even melt.

The most serious problem with foam fluxers is the difficulty in maintaining a sufficiently high head of foam. While it is quite sufficient for clinched or precut leads, it is not easy to obtain an adequate height for long leads or pins. The standard nozzle will produce about ¾ in. (10 mm) of foam above the nozzle top. If a higher head is required, then the nozzle can be fitted with support brushes. These are fixed on either side of the top of the nozzle and, in fact, extend the nozzle upward. The brushes support the foam but allow the long leads or pins to pass between the bristles. However, as mentioned above, brushes also can cause problems, and for longer leads the use of the wave or the spray fluxer is recommended. Extender wings are often fitted to the foam fluxer nozzle (Fig. 5.5). These help to produce a more consistent and slightly higher head of foam but it is not possible to provide more than ½

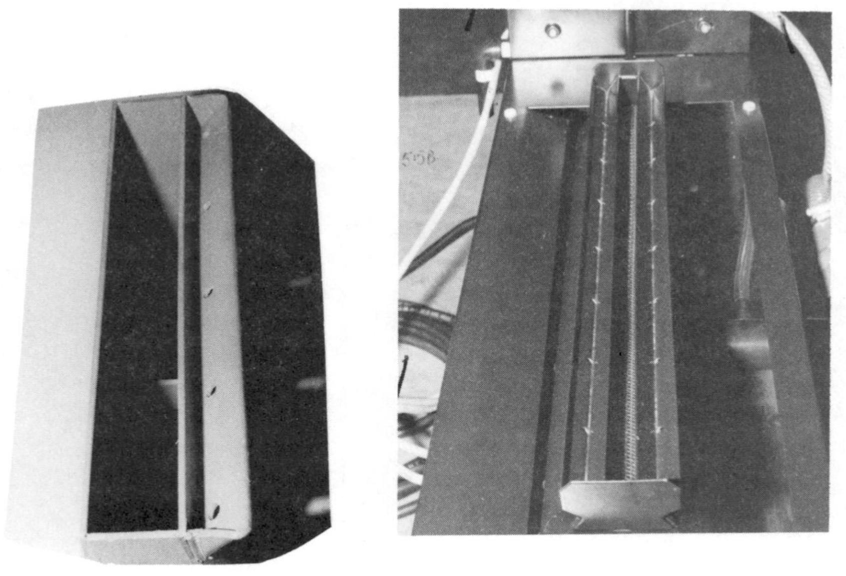

Figure 5.5. Extender wings on two different foam fluxer nozzles. Courtesy Electrovert Ltd.

in. (12 mm). As mentioned, the foam fluxer is an extremely simple piece of equipment. However, there ar several points regarding the operation and maintenance which must be clearly understood and religiously adhered to if a consistently reliable performance is to be maintained.

1. The volume of air supplied to the fluxer must always be adjusted slowly. On initially applying the air, the foam will slowly climb to the top of the nozzle. As the volume of air is increased, the formation of foam will increase and the foam head will rise higher. The foam should be composed of a compact mass of tiny bubbles. If the air volume is now increased further the foam will begin to break up due to the formation of large bubbles. The foam head will intermittently fall as these large bubbles burst. The optimum setting is just before the onset of these larger bubbles. Any increase in air volume beyond this point will only cause a further collapse of the foam (Fig. 5.6).

   There are usually two controls in the air line to a foam fluxer. The first is a pressure reducing valve, which should be set to provide a pressure between 3 and 5 lbs PSI (1.4 and 2.5 g per cm^2). The second is a needle

Figure 5.6. Typical foam fluxer control panel. The top two knobs control the pressure and volume of air supplied to the porous tube. The lower two provide similar control for the air knife. The black knobs control the pressure reduction valves, as indicated on the pressure gauges. The light-colored knobs operate needle valves, which determine the volume of air supplied. Courtesy Electrovert Ltd.

valve to control the volume of air supplied to the porous tube and thus the height of the foam head. All adjustments to the fluxer must be made with the needle valve. If it is not possible to obtain a satisfactory foam head with a low air pressure then the flux is contaminated. It should be dumped and the fluxer thoroughly cleaned with an appropriate solvent followed by a wash in hot water with a detergent and clean water rinse.

2. The very fine pores in the porous tube, or stone as it is sometimes called, are easily blocked by dry flux. The stone must always be kept wet with flux or thinners, and when the fluxer is shut down, it should be stored in flux thinners in a suitable closed container.
3. Flux is usually a sticky messy material. As it dries it becomes difficult to remove. The fluxer, therefore, should always be drained and washed out with thinners if it is to be shut down for more than a short time.
4. Pallets and other fixtures must be cool before going over the foam. If they are hot they will cause the foam head to collapse.
5. The compressed air must be absolutely clean and free from oil. Anything in the air will contaminate the flux. Some fluxers are supplied with their own built-in compressor.

The Wave Fluxer

The wave fluxer is probably one of the older methods of applying flux to the PWB assembly. Once the solder wave had been developed it was logical to use the same system to apply the flux. It consists of a pump and nozzle mounted into a suitable tank or container. The pump forces the flux up through a slot-shaped nozzle, where it forms a wave as it flows back to the tank. The board assembly is conveyed over the wave so that the underside is just immersed in the flux, while the component or upper side remains free of the liquid. The solder side, therefore, is given a thin even coating of flux. The wave fluxer is capable of producing a wave that is deeper than the foam head and is therefore an obvious choice where longer leads have to be fluxed. This method of fluxing also applies more flux than the foam fluxer, and as discussed previously a brush or air knife is normally used to remove the excess flux and prevent dripping onto the preheater (Figs. 5.7 and 5.8). Because the wave fluxer is often used at the higher wave heights and a smooth wave is needed to prevent flux flooding the upper surface of the board, the nozzle is usually designed to incorporate additional elements such as a mesh or perforated screen to reduce ripples and other disturbances. This feature also assists in avoiding unfluxed areas on the board.

The wave fluxer is less critical than the foam fluxer in setup and operation. The wave height is not as dependent as the foam head on the density, temperature, and contamination level of the flux. Hot pallets will not cause

Figure 5.7. Typical wave fluxer. Courtesy Electrovert Ltd.

the wave to be affected in the same way as the foam head. This is, therefore, the fluxer to use in any second soldering station, for example, the solder-cut-solder machine described later in this book.

However, if zero defect soldering is to be achieved, then the same care in operation and maintenance must be observed as with the foam fluxer or any fluxing system. In some ways it is even more important to set rigid standards of control with wave fluxing, since there is no warning of out of tolerence conditions, as indicated by the behavior of the foam head. The pump and pump motor will require maintenance according to the manufacturer's in-

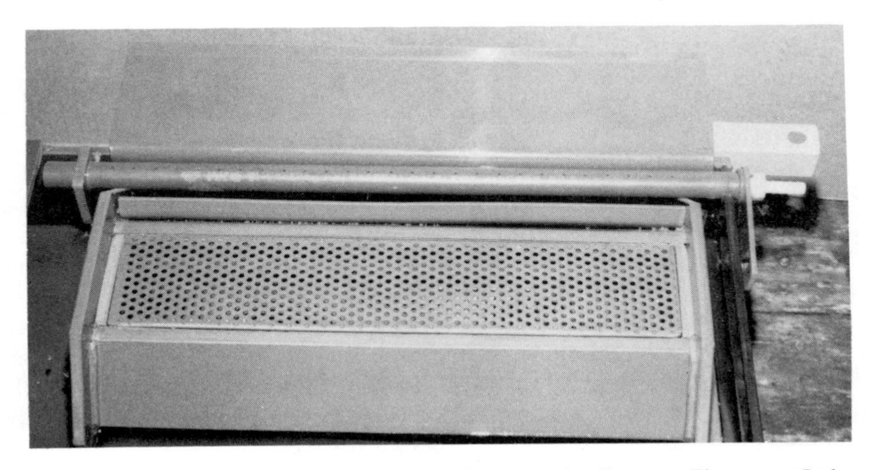

Figure 5.8. An air knife mounted on a wave fluxer nozzle. Courtesy Electrovert Ltd.

structions. As with all fluxers, good housekeeping and cleanliness are of the utmost importance. Fluxes are usually sticky, often corrosive, and any neglect will soon be reflected in the performance of the soldering system.

Spray Fluxers

Spray fluxers are messy, require more maintenance than any other form of fluxer, and inevitably cover the immediate vicinity of the spray with the residual flux. This overspray is a difficult problem to overcome, and therefore spray fluxers are to be avoided if at all possible. However, when long leads have to be fluxed, when an absolute minimum of flux is required, where leads are not crimped, and there cannot be any possible movement of the components, then spray fluxing is the only method available. There are four variations of spray fluxing in use today.

The compressed air spray
The airless spray
The drum and air knife droplet spray
The nozzle droplet spray

Not all these forms of spray fluxer are commercially available; some have been built for in-house use only. Although they all produce a spray of flux, their operating methods and the results obtained are so different that they will all be reviewed in detail.

The Compressed Air Spray The compressed air spray is nothing more than a variant of the normal paint spray, with suitable modifications to enable flux to be used instead of paint. These changes include the correct materials in the nozzles and housings to avoid corrosion by the active ingredients in the flux, accurate controls so that repeatable settings can be made, and the provision in the system for comprehensive protection against overspray. Most of the systems of this type in use today are of in-house manufacture. In fact, there is no spray fluxer of this type on the market. They suffer from nozzle blockage because the air in the gun tends to dry up the flux if the fluxer is out of operation for any length of time, and cleaning then requires taking apart the nozzle assembly.

Nevertheless, this method of applying flux has been made to work and has proved extremely effective for some special requirements. It can be precisely controlled, and therefore the amount of flux applied to the board will be consistent over long periods of operation. In addition, there is no recirculation of the flux; it is a "total loss" system. The flux is drawn from the container fresh and clean and does not see any form of contamination except the compressed air. This means that there is no need to monitor flux density, no addition of thinners to be made, and no need to dump the flux as with recirculatory systems. Thus, there are advantages in the total loss systems, and in the end they are probably no more costly to run with respect to flux usage than other methods of fluxing.

The Airless Spray The airless spray (Fig. 5.9) is a refinement of the compressed air spray and avoids some of the problems associated with that method of fluxing. Instead of using compressed air as the driving medium, the flux is pumped directly through the nozzle at very high pressure. As the flux is forced out of the nozzle it atomizes into an extremely fine mist, without the addition of any air. Because there is no air in the spray nozzle there is not the same problem of nozzle blockage as with the compressed air spray. Also, of course, the absence of the compressed air to propel the flux removes any possibility of contamination from that source. The flux residues will be minimal and almost invisible. Of course, it will flux assemblies with very long leads.

Although they are not common there are several that have been developed in house for some specific purpose.

The Drum and Air Knife Spray The drum and air knife spray is one of the more common forms of spray fluxer and is one of the systems offered for sale by some of the solder machine vendors (Fig. 5.10). This is a totally different concept from the previously described spray fluxers. In this system there is a revolving mesh drum that extends the width of the flux pot. This drum rotates slowly through the flux, and as it does so it picks up flux, which is retained in the openings of the mesh by virtue of its surface tension. At the top of the rotation of the drum an air knife is arranged to blow through the

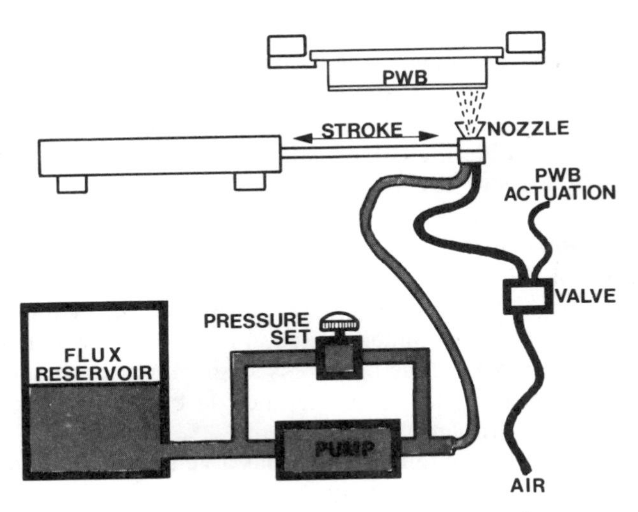

Figure 5.9. A diagram of a typical airless spray fluxer.

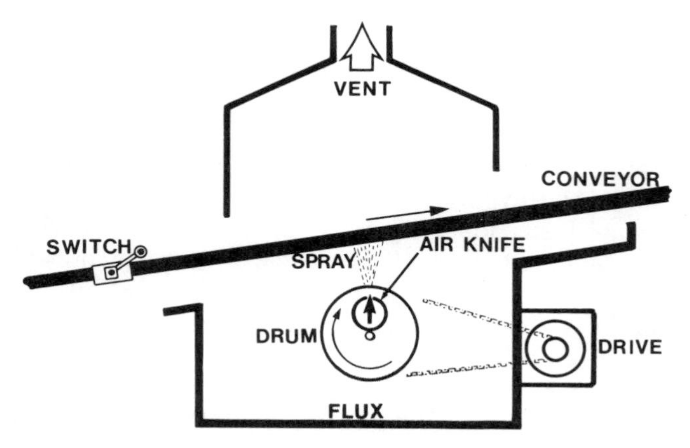

Figure 5.10. A diagram of a typical drum and air knife spray fluxer.

mesh which blasts the flux out of the openings onto the bottom of the PWB assembly. The flux is not atomized in this system but rather is in the form of small droplets. It is an effective form of spray fluxing, and the volume of flux deposited on the board can be controlled to a certain degree by varying the speed of rotation of the drum and the pressure of the air supplied to the air knife. It is not quite as messy as the previous spray fluxing systems, but there is still considerable overspray. Cleaning and maintenance are, if anything, more of a problem because the moving parts of the fluxer and the air knife have to be cleaned regularly. When cleaning, the appropriate solvents must be used, not forgetting that both polar and nonpolar materials may have to be washed away.

Unlike the previously described fluxers, the rotating drum fluxer is a recirculating system; therefore, the same concern for density and contamination control must exist as with the wave and foam fluxers. Contamination is not normally a problem, but some of the spray will return to the pot, and with the open surface of the flux some contamination is always possible. When excellence in soldering is the goal, a tight control must be maintained over all possible areas of concern.

The Nozzle Droplet Spray This development of the spray fluxer has improved on two of the points that are most troublesome with this form of fluxing: overspray and nozzle blockage. In this design of fluxer the compressed air nozzles are surrounded by secondary nozzles which are fed with a supply of flux (Fig. 5.11). The compressed air blows through this primary nozzle, pulling flux from the secondary nozzle cavity and forming a fan-shaped spray of flux droplets.

At the same time the movement of air has a venturi effect and pulls more flux into the nozzle area. In this way a constant spray volume is produced, whose height can be controlled by varying the air pressure in the nozzle.

Figure 5.11. A nozzle droplet spray fluxer. (*a*) An overall view. (*b*) An interior view showing the nozzles. Courtesy Electrovert Ltd.

The ability to control the spray in this way materially reduces the amount of overspray, and in addition the flux is not atomized but stays in the form of droplets. Most of the spray that does not adhere to the undersurface of the board will fall back into the flux container.

There is no flux passing through the air nozzles, and therefore they do not block up when the fluxer is shut down. The nozzles are arranged in a row across the machine at right angles to the line of travel of the assemblies, and each one can be individually controlled or shut off to produce the optimum spray pattern for the boards to be processed.

Although overspray is reduced compared with most other forms of spray fluxers, the same precautions must be employed to contain and control superfluous flux. Maintenance and cleanliness are also important if the fluxing is to be consistent and trouble free.

Spray Fluxer Venting Generally, spray fluxers are not the first choice for applying flux to the assembly before soldering. However, as already mentioned if long leads have to be fluxed there is little option. Also, the ability to apply a minimum amount of flux makes the airless spray fluxer attractive. The fact that some of these forms of fluxing are also "total loss" processes make them a valid option when the cost of automatic flux monitoring and addition cannot be justified, for example, when well-trained or careful operators are not available. The chief objection to any form of spray fluxing, as already mentioned, is the undesirable mess caused by the overspray. This is an unavoidable feature of the system, although the severity of the problem is a measure of the design of the venting of the machine.

All spray fluxers have to be very carefully vented, especially if flammable fluxes or solvents are used where inadequate venting can cause fire or explosion. The solder machine mechanism also requires adequate protection, since the flux can dry out on conveyors, chains, and so on, and if allowed to dry and build up, the flux can cause major mechanical breakdowns.

All spray fluxers must have some form of automatic control so that the spray operates only when there is a board present to be fluxed. The conveyor is usually fitted with some form of position sensor, such as a microswitch, so that the fluxer is triggered "on" when the assembly enters the fluxing station. A timing device or a second sensor is arranged to shut off the fluxer once it has sprayed the bottom of the assembly. If this provision is not included in the soldering machine the problem of overspray is aggravated, and there will be an unnecessarily high flux usage.

Every spray fluxer must be vented independently of the overall venting of the soldering machine and must include some form of flux separator in the venting arrangement (Figs. 5.12 and 5.13). In one typical method the vented air sucks the overspray over a series of flat plates that form a labyrinth mounted in a suitable housing. The droplets or mist of flux will stick to the plates, while the air will pass out of the vent stack. From time to time the plates will have to be removed and scraped clean of the hardened accumula-

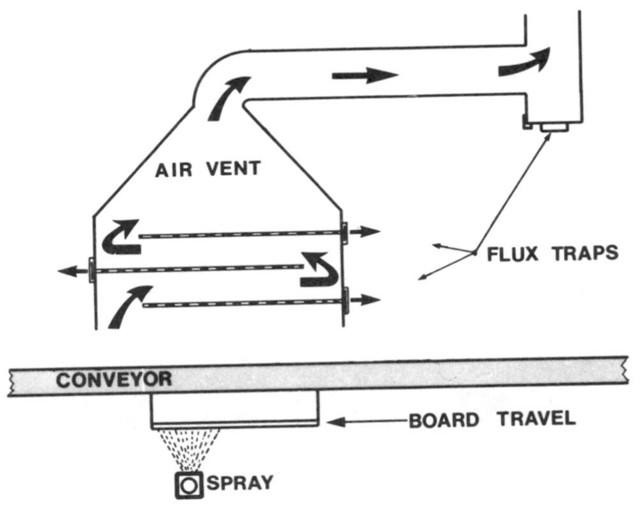

Figure 5.12. A schematic diagram of the overspray protection and flux traps necessary when using a spray fluxer.

Figure 5.13. (*a*) A typical spray fluxer venting system.

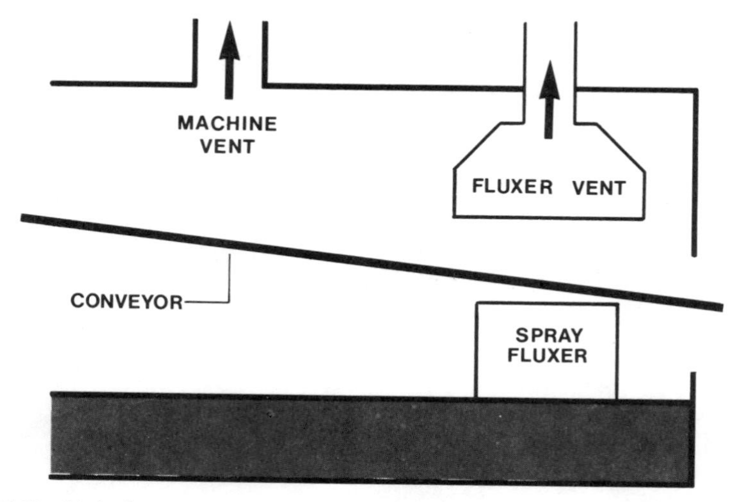

Figure 5.13. (*b*) A diagrammatic view of the separate venting necessary when using a spray fluxer.

tions of flux. Even with this form of separator, however, there will be some flux escape into the duct work, and the design of the ducts must take this into account. The first 10 ft or so must be removable for cleaning, and a suitable trap fitted so that the flux vapors will not condense and run back down to the fluxing station. This ducting will have to be removed and cleaned regularly not only to maintain efficient operation of the soldering process but also to remove a possible fire hazard caused by the accumulated flux deposits. The chemicals found in fluxes are frequently corrosive, and the selection of materials for the duct work must take this into account. The actual materials selected will depend on the fluxes used and should be the subject of consultation with the flux vendor.

The Brush Fluxer

This form of fluxer (Fig. 5.14) is not seen to any extent today, although it did enjoy a period of popularity some years ago. It is a simple method of applying flux and is particularly effective for use with single-sided boards that have the components clinched or otherwise retained. If the components are loose they may be dislodged by the passage over the brush. The fluxer consists of a boxlike container that stretches across the width of the conveyor. In the container revolves a long cylindrical brush that dips into the flux that is contained in the box or flux pot. The brush becomes wetted with the flux and transfers it to the bottom of the board as the conveyor moves the assembly over the rotating brush. It is not easy to control the amount of flux applied to the board. Some variation is possible by varying the depth of the

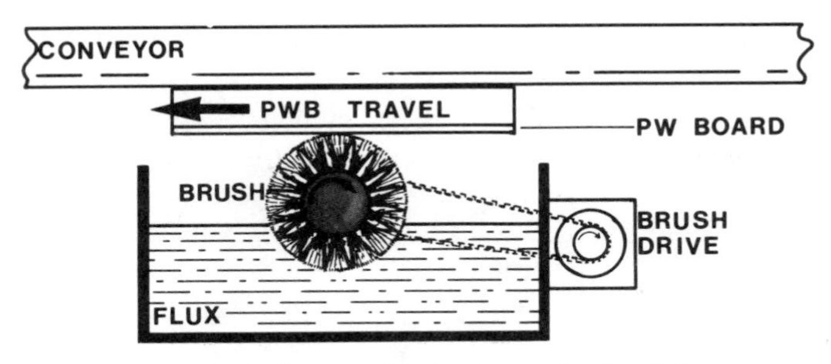

Figure 5.14. A diagram of a simple rotating brush fluxer.

brush in the flux and by changing the speed of rotation. The density and thickness of the bristles have the largest effect, and some experimentation may be necessary to arrive at the optimum brush design. The difficulty in controlling the quantity of flux applied is the worst feature of this fluxing system, but it has certain advantages of simplicity which make it attractive for single-sided board work. It is not recommended for boards with plated through holes, which require a fluxing system that will ensure that the insides of the holes are wetted with flux.

The brush fluxer is a recirculating system and requires control of density and contamination. It is particularly prone to picking up contamination from the assemblies because the brushing action will remove fingerprints, grease, and other foreign materials that may have been inadvertently applied by handling or improperly cleaned parts. This is especially so when used with a flux that has a strong solvent base. The degree of contamination will depend to a great extent on the care in handling during assembly.

The brush must be cleaned thoroughly every time the fluxer is shut down, and it must be stored in a flux solvent when not being used for short periods Otherwise, this is a simple and reliable method of applying flux and should certainly be considered when unsophisticated, single-sided boards are to be soldered.

Operating Rules for Fluxers

With all fluxers, cleanliness and good maintenance are the key factors for reliable operation. Specific details have been discussed already for each individual type; there are, however, several factors that are common to specific groups of fluxers or all fluxers.

1. Except for the "total loss" flux systems, the specific gravity or density of the flux must be checked regularly, and the correct addition of thinners must be made to keep it within the specified limits.
2. As the PWBs pass through the fluxer, in all except the "total loss" types, contaminants from the assemblies will be washed into the flux.

These can reduce the fluxing action, cause the foam to be reduced in height in the foam fluxer, and will eventually cause joint failure. Dump the flux regularly. Begin with every 10,000 ft^2 (920 m^2) of board processed. Log all changes and additions to the flux and note soldering results. Extend the dumping intervals as experience indicates. Cleanliness of parts and boards, good housekeeping, and proper handling can extend the flux life many times.

3. At the onset of any soldering problem, dump the flux, wash out the pot, and refill with fresh flux. This removes one variable from the process and saves time in troubleshooting.
4. Follow precisely the manufacturer's instructions for both the fluxer and the flux. Carry out maintenance as suggested; use only the recommended thinners and cleaning solvents.
5. Never mix fluxes and do not change from one flux type to another without carrying out tests.
6. Even changing flux vendors should be done with caution; there are variations in flux formulations even with fluxes to the same specification.
7. A careful record must be maintained of the fluxing process—flux changes, additions, density measurements, thinner additions, setup parameters, and so on. As will be seen in later chapters, statistics, correctly maintained, are essential for efficient operation.

In summary, the choice of a fluxing system is dependent on the construction of the boards to be soldered and the method of assembly. If the leads are to be clinched or precut, so that there will be very little protrusion below the bottom of the board, a maximum of ⅜ in. (10 mm), then the foam fluxer is the best choice—simple and low cost, but sensitive to flux contamination and hot pallets.

Where more flux is required on the board, holes require full fluxing, or for some other reason it is not possible to use a foam fluxer, then the wave fluxer is the next choice. The maximum wave height is ½ in. (12 mm), unless a special deep nozzle is used for which a wave of about 1 in. (25 mm) can be generated, although not without some turbulence. But again this is a comparatively simple system, which is not temperature sensitive and does not require a supply of compressed air as does the foam fluxer.

When the assembly has long leads, the first choice should be the foam fluxer with support brushes to give a higher head of foam. If this is not adequate, or if the leads must not be disturbed, then the only alternative is to go to the spray fluxer. This is an effective, but messy, answer to the problem of fluxing long leads. Before making a decision on the type to acquire try those available on some actual production boards. Check the soldering results, check the amount of overspray, and thoroughly investigate the methods of control built into the soldering machine. This is a vital part of the spray fluxing system.

Flux Density Control

When the volume of production is high, the measurement of flux density and the necessary additions of thinners can become a considerable chore for the operator and, in the rush of production, can be easily overlooked. Some form of automation, therefore, is almost a necessity under these conditions, and there are several systems of varying degrees of complexity available.

The initial problem of maintaining the level of flux in the fluxer is solved for the lower throughputs by means of the bottle feeder (Fig. 5.15). As the level of the flux falls, the mouth of the bottle is exposed, and the inrush of air allows flux to escape until the mouth is once more covered and a partial vacuum formed in the bottle, preventing any more flux flowing out. This is a very old principle and is used in many domestic applications—a simple but effective device. However, this still requires the operator to make regular checks on the flux remaining in the bottle.

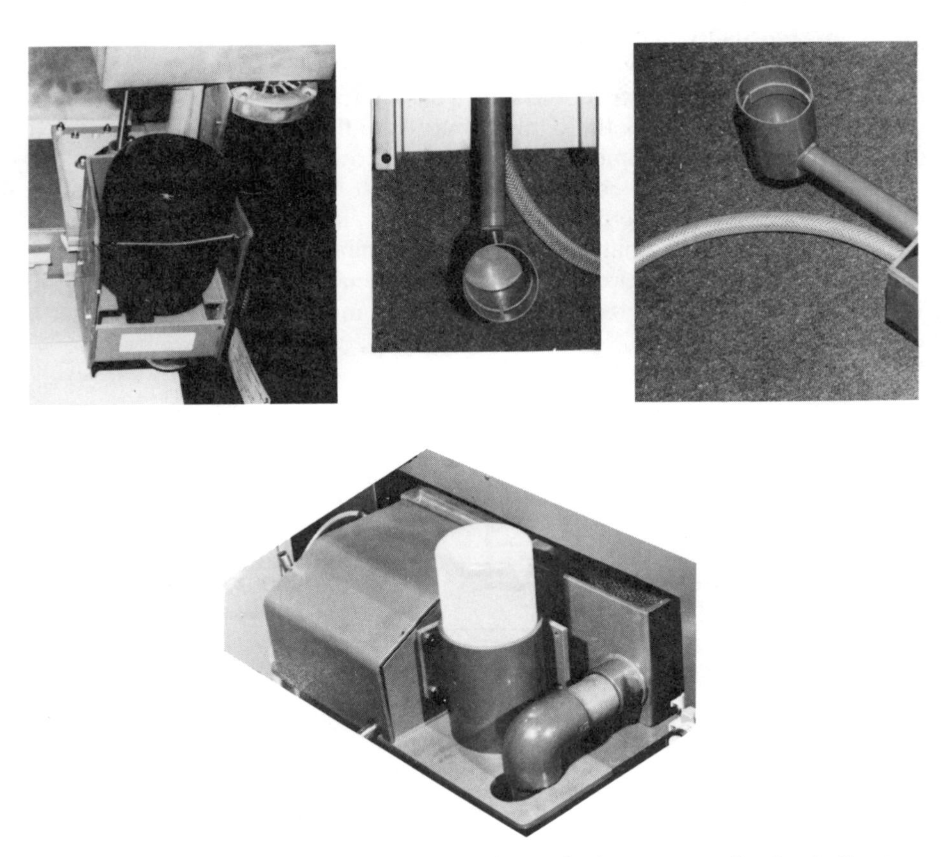

Figure 5.15. Simple inverted bottle feeders used to maintain a constant flux level. Courtesy Electrovert Ltd and Zevatron GMBH.

A longer-term solution to maintaining the flux level uses a constantly running pump and is shown in diagrammatic form in Fig. 5.16. The fluxer contains the minimum amount of flux necessary for efficient operation of the system, while the main volume is contained in a tank remote from the fluxer itself. The two are connected by means of two pipes and a small pump. The larger of the two pipes connects to a simple weir in the fluxer which determines the height of the flux level. This weir can be just a simple connection to the pipe, or it may be more complicated and include some form of height adjustment. The second pipe is considerably smaller in diameter and is connected to the pump, which continuously adds flux to the fluxer from the main tank. It is obvious that the outlet pipe must be large enough to accept all of the overflowing flux, no matter how fast the pump operates. In practice, the pump is set to do little more than supply a trickle of flux.

The flux, therefore, will continue to circulate until the main tank is empty, and during this time the fluxer will always have the same level of flux, determined by the height set by the weir.

As well as being a simple system, this does not require an electrical or other level measurement device, and the continuous pumping will keep the flux mixed. Suitable valves can be fitted so that when the system is shut off the flux will drain down from the fluxer back into the main tank. Of course, with this system there is the danger that if the flux becomes contaminated accidently, a larger volume will have to be thrown away.

There are other level control systems which rely on sensors to monitor the amount of flux in the fluxer and operate a pump to add more as necessary. These can use a float whose position is measured by magnetic capacitive or other means, or probes that are activated by the conductivity of the flux. All operate satisfactorily, some will have more hysteresis than others,

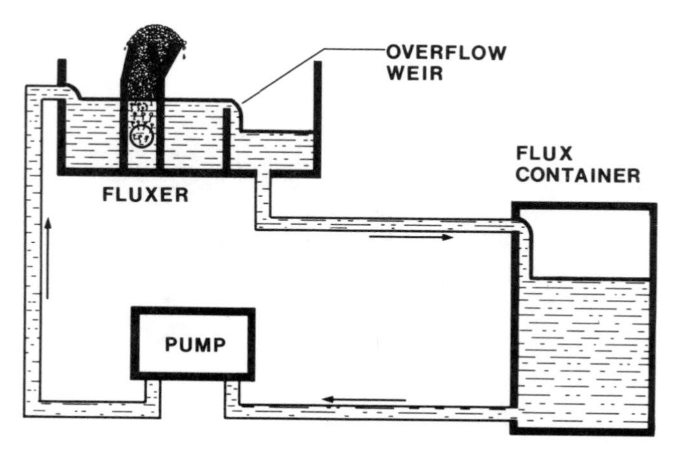

Figure 5.16. A diagram of a pumped flux system used to maintain a constant flux level.

some will be more sensitive, but in practice these are not major differences. The main problem with most of them is the messy nature of most fluxes, which can cause sticking if the systems are not maintained according to the manufacturer's instructions. Variations of the flux level are not critical. The foam fluxer is probably the most affected, although variations of ¼ in: (6 mm) are acceptable. The usual symptom of low flux is the appearance of large bubbles, which require the air pressure to be reduced in order to obtain a fine foam that does not collapse. This, in turn, results in a lower foam height. These effects are very dependent on the type of flux used.

Density control of the flux is a much more complicated matter. As discussed previously, this must be controlled very carefully if the process is to operate reliably. The simplest method, of course, is to use an ordinary hydrometer, and with a conscientious operator, who is given time to carry out this function, it can be reasonably satisfactory. However, the addition of the right amount of thinners to bring the flux back to the correct density is not quite as simple.

It is strongly recommended that the following formula be used to generate a table giving the amount of thinners to be added in order to maintain the correct flux density. This avoids the possibility of overcorrecting when making additions.

$$V = C(m - d) - (d - t)$$

when, for a given temperature,

C = the capacity of the fluxer in cubic centimeters
V = cubic centimeters of thinners required
M = measured specific gravity of flux
d = desired specific gravity of flux
t = specified specific gravity of thinners

It is possible to make an approximate correction for temperature by adding or subtracting 0.001 for every degree Celsius above or below respectively the temperature at which the specific gravity is specified.

In addition to measuring the density and making the additions, these data must be recorded accurately. A typical flux control sheet is shown in Fig. 5.17 or it may be combined with the daily record. With most machine soldering operations the density should be checked after every 2 hours of operation. The records will quickly show if the measurements need to be made more frequently or if the measuring frequency can be reduced. Because there are so many variables in the process, this should be discussed with the flux vendor, whose recommendations should be used as the basis of the operating procedure. Obviously, this control can become a major part of the operator's task if carried out correctly, and therefore any automation will be economically justified.

FLUX CONTROL RECORD				MACHINE No. ________		
FLUX TYPE ________		SG ________		THINNERS ________		
Date	Time	SG	Additions	Operator	QC	Comments

Figure 5.17. A typical flux control log.

Automatic Density Controllers

There are now on the market many systems that will carry out the function of automatic density control (Fig. 5.18). They all incorporate some form of automatic level control in the package. One popular method of achieving this is the constantly pumped system described above. When a sensor is added to this to measure density of the flux, the pump can be signaled to continue to provide flux or to pump thinners when the density rises. In some systems, this is done by using two pumps, one that is constantly circulating the flux and one that pumps only thinners on demand from the sensor. In other equipment only one pump is used, and the sensor operates solenoid valves that allow it to add flux or thinners as required. The density sensor has to be quite sensitive to record small changes in the flux and operates very much in the same way as the float level sensor, except that now, of course, it is working with a constant level of flux, and therefore any changes in the density of the liquid will be reflected in the level at which the sensor floats.

One very effective system measures viscosity rather than density. The sensor is completely enclosed and wetted with flux. As a consequence it does not suffer from the sticking problems frequently found with float systems (Fig. 5.19).

There is a danger that when small changes in density are sensed, the pump may continue to add thinners until the effect is read by the sensor, by which time so much thinner has been added that the density has been made too low. In order to avoid this most systems operate on a timed sequence. In this method when the sensor indicates that thinners are required, the pump is timed on for a short period and adds a small amount of thinners. It then shuts down while the mixture is evenly distributed throughout the system. A second measurement is then made, and so on until the sensor indicates that the

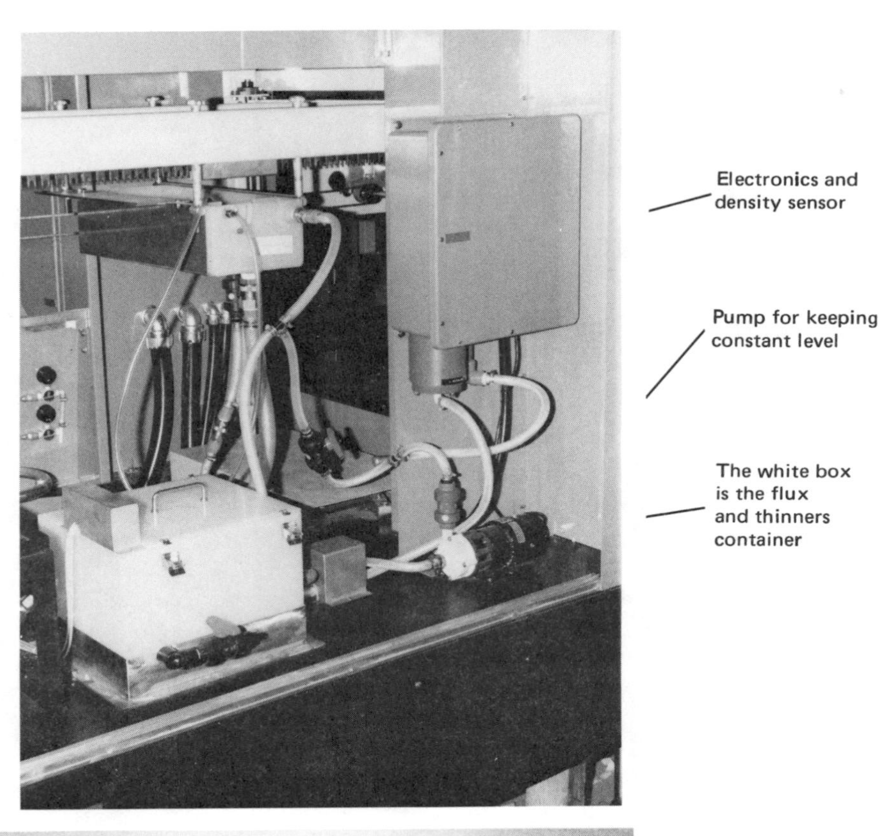

Figure 5.18. An example of an automatic flux density control system. Courtesy Electrovert Ltd.

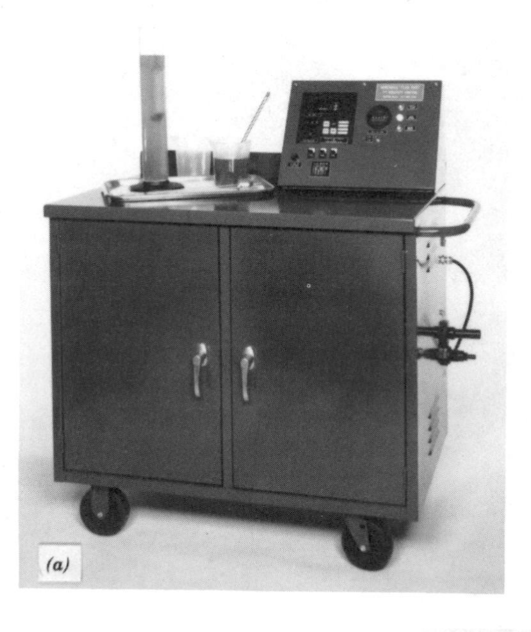

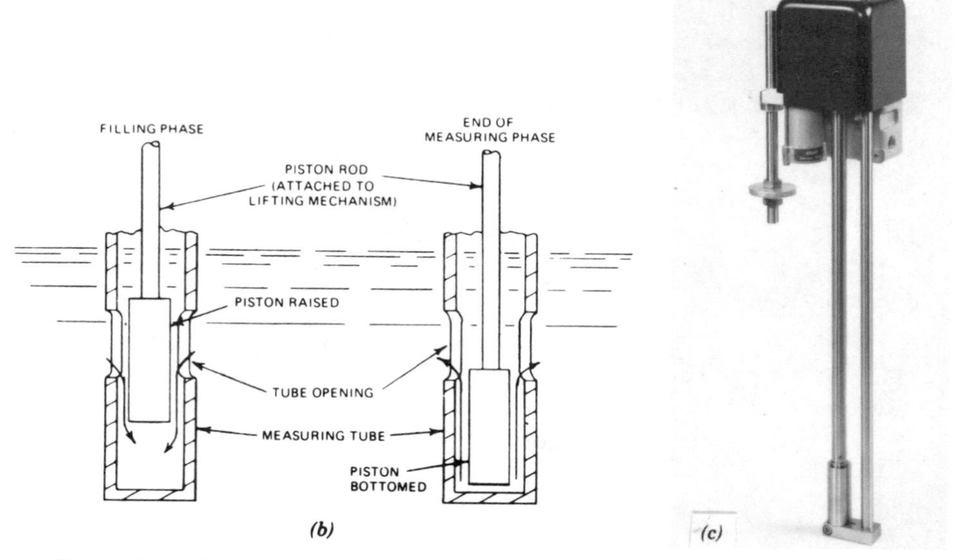

Figure 5.19. (*a*) A complete flux density control system using viscosity measurement as the control factor. Courtesy Norcross Corp. (*b*) Principle of the classic viscosity measurement using a falling piston. Courtesy Norcross Corp. (*c*) Picture of a typical viscosity transducer.

density is within the specified limits. This also means that the overall density figures will be held to a tighter tolerance.

Obviously, to control the density accurately temperature has to be considered. In most of the automatic control systems there is a temperature sensor mounted in the flux, and the controller automatically computes the temperature compensation required. Some of the more simple density controllers do not have this feature, and while they do not achieve the optimum accuracy, in practice they hold the flux density within acceptable limits and are very attractive because of their low cost and simplicity.

While there will be quite a temperature swing in the flux when the soldering machine is switched on, once it is running the nearness of the preheater to the fluxer and the general effect of the machine venting will be found to hold the flux temperature remarkably constant. The Mil Specs require the flux density to be held within plus or minus 2%, and this is within the capability of any of the controllers on the market today.

Flux density is important in arriving at zero defect soldering. Almost any automatic control will do a better job than relying on the operator with a hydrometer. It is just too easy to forget, to make a mistake, or to get too busy with other things.

With all density measurement and control systems, from the humble hydrometer to the most complex controller, a regular quality control (QC) check of the accuracy of the equipment must be part of the overall control and the results included in the flux control records. When selecting an automatic flux density and level control system, the following features should be considered.

Can the system be cleaned easily?

Will the pumps and sensors clog up if the system is shut down for a short while?

What are the long-term and short-term accuracies?

Will the system operate reliably on a day-to-day basis without making any adjustments?

Is there adequate warning of failure?

Will the controlled density cover all the fluxes likely to be used?

Can the density set point be changed easily?

Will the materials used in the construction of the system stand up to the chemicals in the fluxes to be used?

In the event of failure of the controller or sensor, can the fluxer be operated manually?

Without doubt the most important feature is the reliability and the ease of cleaning out the pipes, sensors, and so on. Before purchasing any system try to see one in operation in a production environment. The layout of a typical system is shown in Fig. 5.20.

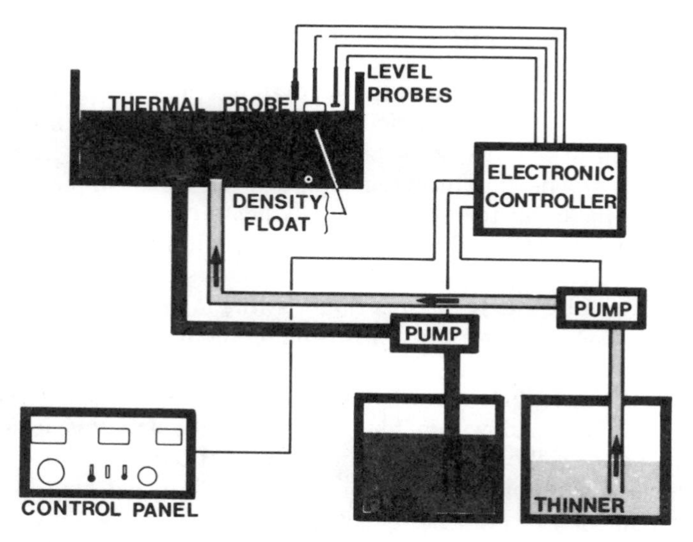

Figure 5.20. A schematic of a typical float sensing flux density control system.

PREHEATERS

Preheating is the practice of applying heat to the printed wiring assembly after fluxing and immediately prior to soldering. Why preheat? It is not mentioned in the requirements for a good soldered joint. The reason is that preheat is not necessary in order to make a sound soldered joint. Soldering can and is being carried on without it. However, it was noted earlier that there is a requirement for "the correct temperature for the correct time." In mass soldering the correct time is achieved in two stages: the first is preheat, and the second is the solder wave.

Without preheat, soldering can be carried out and will give excellent results. However, the soldering speed will be extremely slow. In some tests carried out during experiments on preheating, a board was soldered at 12 ft (4 m) per minute with preheat. Without preheat it could not be soldered faster than 2.5 ft (0.7 m) per minute. The very essence of mass soldering is to solder economically, and this infers quickly and reliably. In this sense, preheat is absolutely vital and is, in fact, one of the more important parts of the soldering process.

In applying heat to the board assembly every attempt is made to heat the metals that are to be joined, while keeping the heat away from the rest of the board and the components. This is seldom achieved, and it will become apparent as this chapter progresses that even the best preheat system is very much a compromise.

The usual reasons given for using preheat are as follows:

To reduce the thermal shock to the board from the solder wave

To "activate" the flux by heating

To dry off flux solvents, which can cause blowholes in the solder joint by boiling during soldering

Now the Mil Specs require that a PWB must accept a 10 second float on a solder bath without damage to the laminate or cracking of the copper. The board will see considerably less exposure than this to the solder in its passage through the soldering machine. Therefore, the premise that preheat is used to prevent this mysterious solder shock must be questioned. Leaded components will normally only have the leads contacting the solder and again the idea of damage from thermal shock cannot be taken seriously.

Certain surface mounted chip components pass through the solder wave and can be damaged by the sudden increase in temperature from contact with the molten solder. This particularly applies to certain chip capacitors. In these cases the preheat temperature and the rate of preheat can be important in limiting any damage. For example, one capacitor manufacturer limits the rate of heating to 1°C per second and the maximum temperature between the preheated board and the solder wave to 100°C.

In a similar vein a review of flux properties will show that the temperature reached during preheat will not materially increase the chemical activity of many fluxes. Certainly most of the volatiles must be removed prior to the solder wave, but this is not a problem unless the flux is heavily applied.

The most important reason for using preheat is always left out; this is the need to heat the PWB, the component leads, and all the other parts of the joint so that the final heating and soldering can proceed more quickly. In some cases the actual exposure to the hot solder has to be minimized, and then preheat becomes even more necessary and critical.

For example, when chip components are to be soldered there is a limit to the length of time that the chips can stay in the solder. The metallizing that provides the electrical connections to the ends of the devices can be dissolved by the solder, and prolonged exposure to the wave results in parts that have been stripped of their connections. In this case the assembly must have the maximum amount of preheat so that the soldering time is reduced to the minimum. As assemblies become more densely packaged, the thermal capacity increases, and preheat becomes more and more important in achieving the required soldering speeds. With multilayer boards there is a special problem in obtaining sufficient preheat quickly enough. This has caused the use of top preheaters as well as the more normal system for heating the bottom of the board. Even with this additional heat, boards with many layers, and those with several ground or voltage planes, will inevitably have to be soldered more slowly than a similar sized single- or double-sided board. The reason, of course, is the heat sinking effect of the masses of copper in the multilayer boards. Even with the additional preheat much more heating will have to be supplied by the wave, or the board will have to sit much longer over the preheater, or more probably a little of both. Remember, in Chapter 1 the point was made that soldering cannot occur until the mass of the joint is up to the melting temperature of solder.

Types of Preheaters

There are many different ways of heating the assembly prior to soldering. Obviously any method of generating heat can be used, but there are certain factors that limit the choice, and the following are the most commonly used forms of preheaters.

- The electrically heated hot plate or platen
- Tubular electric heaters or cal rods
- Heated air
- Quartz plate electric heaters
- Quartz rod electric heaters
- Infrared electric lamp-type heaters

There is no "best" form of preheating; all of these methods have their advantages and disadvantages and they will all be reviewed in turn so that the characteristics of each can be determined, for helping in the correct selection for any particular requirement. The ideal preheater has not been developed. When it is it will have the following specifications.

- It will have a small thermal mass resulting in very fast heat up and cool down.
- It will use energy efficiently.
- Cleaning will be easy, ideally it will be self-cleaning. In any case dirt or flux contamination must not affect the thermal output.
- It must heat the metal of the joint without raising the temperature of the base laminate.
- The heating must be even all over the board.
- The heating width must be adjustable to accommodate varying widths of assembly.
- The thermal output must not vary with age.

There is no preheater on the market today that complies with all these requirements. There are preheaters that will carry out some of these functions, there are a few that approach the ideal, but generally it is necessary to choose the type that will be adequate for the particular work to be carried out, even if it is not perfect.

The Hot Plate The hot plate is an extremely simple and low-cost method of heating the PW assembly. It is virtually self-cleaning, since any buildup of residues will burn off and will not in any case affect the heating capabilities. If for any reason it does become necessary to clean the hot plate, then scraping with a spatula is all that will be necessary.

This preheater consists of a metal plate, approximately 1/4 in. (6 mm) thick, with strip heaters bolted to the underside. The heaters are usually mounted parallel to the line of travel of the conveyor so that sections of the heaters, if required, can be switched off when narrow boards are being processed. However, the plate has to be sufficiently thick to provide an even heating so that this method of saving power is at best a compromise. The edge heaters are frequently on separate controllers so that the heat losses at the edge of the plate can be compensated for by adjusting them separately from the center heaters. These types of heaters will provide adequate energy for all normal soldering. However, they will not provide sufficient heat for boards with a very high thermal mass unless two or more are mounted in series to provide a longer heating exposure for the boards. This usually results is a preheating section of about 4 ft (1.2 m) in length.

This system of preheating is not a particularly efficient user of energy, but the chief objection to the hot plate is its extremely high thermal mass. This is necessary to produce the even heating, but in consequence it requires a long time to reach operating temperature from cold and a considerable time to make anything more than a very small change in temperature. Depending on the type, initial heating can take up to 45 minutes, and small changes up to 15 minutes.

This is not an inconvenience if many boards are to be run at the same temperature, but it can reduce the output of the soldering operation if many different board types are processed and frequent changes to the preheat temperature become necessary (Fig. 5.21). With some types of preheaters

Figure 5.21. A hot plate–type of preheater. Courtesy Electrovert Ltd.

aluminum kitchen foil is used to prevent them from becoming soiled by flux droppings. It should never be used for this purpose with hot plates. The flux will not harm the surface, as mentioned above, and the shielding effect of the foil will damage the heaters and reduce the output of the hot plate. Similarly, no attempt should be made to use heatproof paint with the idea of protecting the surface.

Cal Rod Heaters The cal rod preheater is faster in response time than the hot plate. It is probably no more efficient in its use of energy. It consists of tubular elements that operate at or near red heat. They are fitted into a frame that usually contains some form of reflector to assist in transmitting the infrared energy up to the bottom of the board. The cal rod heater, therefore, is to some extent dependent on the effect of the reflector, and there may be some variation in the heat output as the reflecting surface becomes contaminated with burnt flux. A common procedure is to line the reflector with aluminum kitchen foil and discard this for a fresh sheet when it becomes discolored. The cal rods themselves operate at such a temperature that they are self-cleaning. The elements can be arranged in line with the conveyor and therefore produce variable width heating. The heating is reasonably even over the whole preheater, if the reflectors are kept clean. The cal rod preheater usually operates at a higher temperature than the hot plate and may not require such a long preheater to arrive at the same board temperature (Fig. 5.22).

Figure 5.22. A typical cal rod preheater. Courtesy Zevatron GMBH.

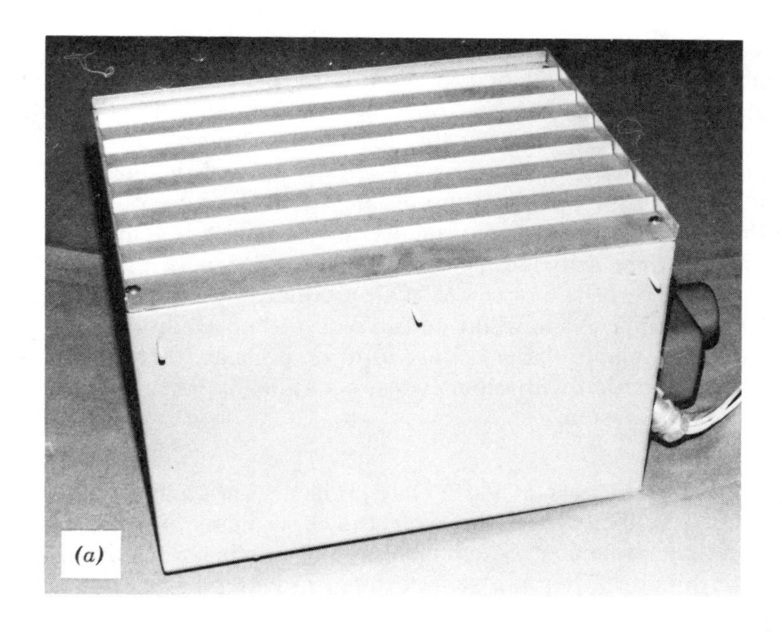

Figure 5.23. Two typical hot air preheaters. (*a*) Courtesy Electrovert Ltd. (*b*) Courtesy Hollis Engineering.

Hot Air Preheaters A hot air preheater was probably the first form of preheating and was developed to assist in drying the then commonly used alcohol rosin fluxes (Fig. 5.23). As a method of getting rid of the alcohol it works extremely well, and for this reason is still used in many soldering machines. As a preheater to raise the temperature of the board it is almost useless. Heating the air is not the problem, but the rate of transfer of heat from the air to the board is extremely slow. The rise in temperature of the board in the time available is very small compared to other methods of preheat. However, the movement of air produced by the hot air preheater is useful in assisting evaporation and avoids the possibility of a dangerous buildup of flammable vapors. This form of preheat, therefore, is usually found in combination with other types, occasionally being built as part of a plate or cal rod system.

Quartz Plate Preheaters Quartz plate preheaters are not as common as the previously described types. However, they have many of the characteristics of the perfect preheater. Their heating and cooling rates are fast, 200°F (93°C) per minute. They are self-cleaning, provide very even heating, and can be arranged in sections to provide a variable width of heated area. There are, of course, some negative features. This form of heater is more costly than the systems already described. It requires more complex control circuits because of the fast response time, and it can be permanently damaged by a comparatively short period of over temperature operation. It is also much more easily damaged by careless handling or by dropping a heavy object onto the surface. However, it is becoming more commonly used because of the need for more preheat on complex boards with high thermal mass, especially in the case of multilayer boards. It operates at higher temperatures than the previous preheaters, and therefore adequate heating can be obtained without having to go to unnecessarily long systems, which are difficult to fit into the standard soldering machines.

As its name suggests the quartz plate preheater consists of a thick quartz plate with heating elements embedded into the material in suitable pockets or slots. Also included are spaces for the controlling thermocouples. Much of the energy from this type of preheater is in the form of infrared energy in the 3 μm band, with a small amount in the form of convection heating (Fig. 5.24).

Quartz Tube Preheaters The quartz tube preheater is very similar to the previously described system and is even faster with respect to cooling and heating. The elements in this form of heater are mounted in quartz tubes, together with the controlling thermocouples. The assembly of heating tubes is more fragile than the quartz plate heater but does offer some advantages. The rods can be mounted parallel with the conveyor and will provide an almost infinite control of heated width by switching each tube separately.

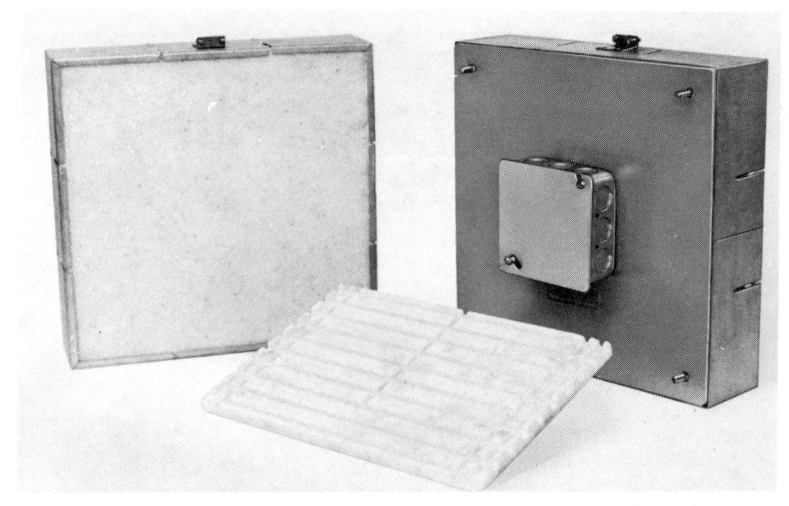

Figure 5.24. A quartz plate preheater. Courtesy Solar Casso Inc.

Alternatively, they can be mounted across the conveyor and individually switched to produce the required amount of preheat, while reducing the energy use under minimum heating requirements. In addition, hot air can be blown through the tubes to provide the advantages of mixed infrared and hot air preheat in the length of a single heating element. If required, ambient air can be used to obtain heat from the quartz tubes. Because of the advantages described, it is likely that the quartz tube heaters will become more popular in the future in spite of their cost and fragility. With the introduction of more automation into the soldering machine, the flexibility, speed of control, and ability to pack a great deal of heating power into a small area make this form of preheating a natural choice (Fig. 5.25).

Lamp-Type Infrared Heaters Lamp-type infrared heaters have the fastest response time of all. With only the lamp filament as the thermal mass, the heating and cooling times are measured in seconds. Peak temperatures that can be reached are extremely high and excellent control is necessary to avoid damage to the board. The lamps are expensive, often have gold plated reflectors that may require water cooling, and must be protected from dirt and flux since they are virtually impossible to clean without scratching and causing permanent damage. With all these disadvantages it is understandable that few machine vendor offers this form of preheating as a standard feature.

When very fast heating is necessary, however, or higher than normal temperatures or fast changes to the preheat are required, then this is often the only system that will solve the problem. Lamp-type infrared heating

Figure 5.25. An example of a quartz tube preheater. Courtesy Electrovert Ltd.

therefore is not uncommon in special machines and in-house designed systems for very special purposes (Fig. 5.26).

Heating, Wavelength, and Infrared

In preheating, the object is to heat the metallic structure of the joint, that is, the connecting wires, mounting pad, and through hole, but not the laminate from which the PWB is fabricated. To achieve this would require that all the energy radiated from the heater be absorbed by the metals and totally reflected by the epoxy glass, paper phenolic, or other insulating materials. In practice this does not happen; indeed, exactly the reverse is usually the case. The properties of the materials are such that the shiny surfaces of the circuit pads and the connecting wires will reflect far more of the energy than the insulating surface of the board.

This is one of the problems of using the lamp-type infrared preheaters. The short wavelength of these emissions, about 1.5 μm, can burn the epoxy glass surface while the copper or solder coated metals will not be heated to any great extent. Actual burning, of course, must be prevented by correct adjustment of exposure time, lamp focus, and lamp temperature. This problem can be understood if it is remembered that the reflectors used with the lamps themselves to collect and focus the energy are made of a shiny reflective metal. While some cooling of these reflectors is usually necessary, they

Figure 5.26. Three lamp-type preheating elements.

obviously reflect the larger part of the energy generated by the filament of the lamp.

In fact, the lamp-type infrared preheating system was quite popular in the early days of soldering but was superseded by other methods of heating when the gold plated finish of PWB conductors became commonplace. The difficulty of heating with the short wavelength radiation made users turn to the use of the hot plate, cal rods, and heaters with radiation in the 5 μm region. Of course, cost was also a major feature in this trend.

As the wavelength increases, the heating of the insulating base and the metallic circuitry becomes more even, but the ability to focus the energy and achieve a fast response time are lost. It is extremely difficult to lay down any hard and fast rules on the subject. The emissivity of surfaces, that is, their ability to absorb or emit energy of different wavelengths, depends on many factors (color, texture, and material) and cannot be judged by visual appearance. No matter which of the heating systems are used, there will always be some convection heating, and the radiated energy can vary in wavelength with some other factors, such as the voltage on the heaters, the age of the heaters, and in some cases the cleanliness of the heater surface.

With the exception of the lamp-type infrared preheater, the problems of emissivity and wavelength will not prove to be a problem in practice, at least where commercial heating systems are used. However, an understanding of these fundamental principles are necessary to be able to comprehend what is taking place in the preheating station, to develop preheaters for specific purposes, and to troubleshoot when problems arise.

Automation in Preheating

As has been discussed, the temperature of the preheater has little comparison with the actual temperature of the PWB. Wavelength, emissivity, conveyor speed, board design, and many other lesser factors will all produce different board temperatures from a preheater operating at a constant temperature Therefore, it can be seen that the measurement of the preheater temperature which is usually the controlling factor in the preheating station, can at best be only an approximation of the control required to maintain a constant board temperature at the solder wave. It is equally pointless to measure the board temperature after preheat, unless the soldering time can be adjusted accordingly, and this is not easy to do at the high throughput rates necessary for efficient soldering. If the various parameters could all be controlled or measured, then the preheater could be modulated to compensate; but, here again, some factors cannot be controlled, for example, the emissivity of the PWB or changes in its thermal capacity.

The answer lies in a computer controlled two-stage preheating system (Fig. 5.27) that is now available. In this equipment the temperature of the board is measured by a remote sensing infrared detector after it has passed over a section of the preheater. A microprocessor calculates from this temperature the amount of additional heat necessary to produce the required temperature at the surface of the PWB by the time it reaches the solder wave and adjusts a fast response heater accordingly. All the factors mentioned above are accounted for in the calculation, and the only input required is to set the final board temperature. This is not a simple or low cost system, but it eliminates another variable from the soldering process.

Selecting Preheaters

With the exception of the automatic system described above, there is little difference in the actual performance of the various forms of preheating. The choice, therefore, is largely one of convenience, and the ability to provide adequate heat to raise the assembly to the required temperature at the maximum conveyor speed. In some machines the space alotted to the preheating stage is not long enough to use the simpler forms of heating and this may require the use of the more expensive, but higher temperature, quartz heaters.

Tests should be run with assemblies that have the largest thermal mass likely to be processed. Run them at the maximum conveyor speed that is likely to be used, and be sure that the preheaters will bring the boards to the surface temperature necessary. Allow for some reduction in output from the heaters because of age or low supply voltages.

Otherwise choose the simplest heater that will do the job and require the least cleaning or maintenance. All preheaters require adequate controls so that once a setting has been established it can be repeated accurately. These are discussed in detail in the section on controls.

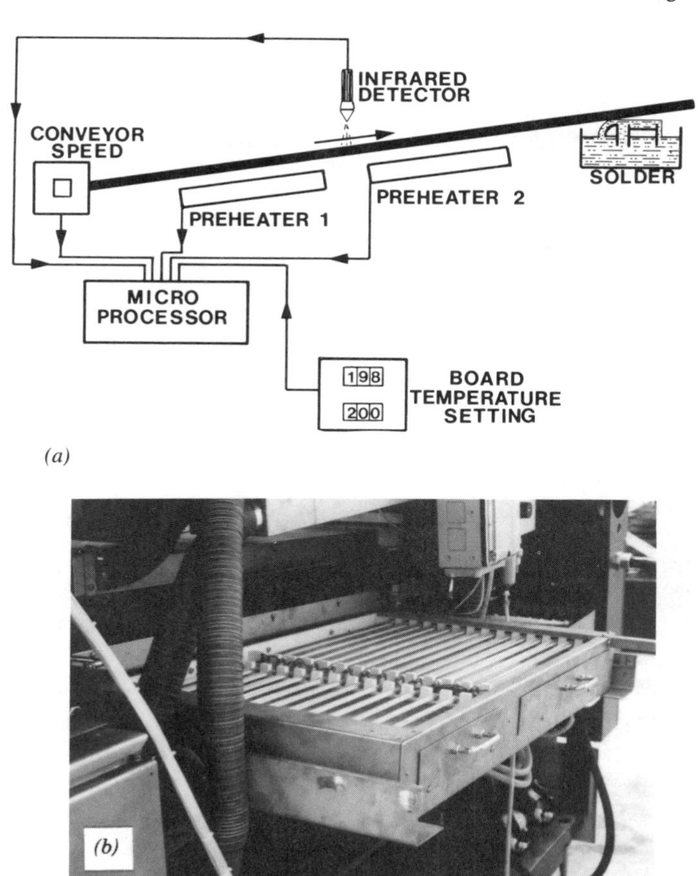

Figure 5.27. (*a*) A schematic diagram of an automatic preheating system. (*b*) A picture of the two sections of a computer controlled automatic preheating system. Courtesy Electrovert Ltd.

THE SOLDERING STATION

The soldering station is the heart of the soldering machine; it contains the solder pot, the pump, and the nozzle that forms the actual wave. While the modern soldering system is extremely reliable, it has to be serviced and cleaned regularly, and the machine must be designed with this in mind. Some soldering systems have integral roll out solder pots, which make maintenance very easy; others arrange for the covers to be rolled or lifted clear, and one incorporates an ingenious elevator that raises the entire superstructure of the machine leaving the solder pot clear to be worked on.

There are many different variations in the design of the actual pumping chamber of the solder pot, but they almost all fall into two categories. In the first, the pump is connected to the nozzle by some form of pipe or channel that is removable from the pot together with the nozzle mounting plate and the actual pumping chamber. This allows easy access to the pot for cleaning once the solder is drained away. However, unless the connection between the pump and nozzle is extremely generous it can limit the volume of solder that can be pumped. The second system divides the pot into two chambers, with the upper joined to the lower by the pump orifice and the nozzle. The solder flow here is limited only by the nozzle and pump capacity, and large volumes of solder can be pumped at low pressures to produce very deep waves, and very wide waves, with extremely smooth turbulent-free surfaces.

The Solder Pot

The container in which the solder is melted has to withstand severe environmental conditions. First, it must withstand the temperature of the molten metal, which with a typical alloy for electronics use will be in the region of 500°F (260°C). With other alloys this can be as high as 900°F (485°C). The solder is heavy, and the solder pot must be strong enough and adequately supported so that the weight will not cause it to distort in any way. In addition, the solder will attack many metals, and the flux and flux residues may be chemically active and cause corrosion.

The choice of materials for the pot therefore is limited, and in practice, cast iron and stainless steel are the two metals most frequently used for this purpose. Both have their advantages and disadvantages, but these are chiefly in the field of fabricating the pot; in use there is little to choose between them. Some more exotic materials are occasionally used for special pots, for example, titanium. A few manufacturers use mild steel with a protective coating that has to be renewed occasionally, about once every year of use.

Where temperatures above 600°F (315°C) are used it is customary to give the inner surface of the pot a coating of oxide or other treatment to prevent attack by the solder until the metal generates its own oxide protection, which occurs quite rapidly in use. Corrosion or other attack of the pot is rarely a problem, especially with the lower melting point alloys. Whenever corrosion has been a problem, it has generally been found to be caused by some material that has been introduced into the pot during the soldering process. In one case of a high-temperature pot used for stripping and tinning enameled wire, it was found to be caused by the residues from the wire insulation. Even where high temperatures are used with very corrosive fluxes the pots have operated for several years without any failure.

Of course, much depends on the flux used, the frequency of cleaning, and the contaminants that get into the pot; no firm rules can be made regarding pot life. As has been noted earlier the materials generally used for solder pots generate a protective oxide coating, which prevents attack by the solder.

When pots are cleaned, therefore, it should be done with care and no tools, such as wire brushes, files, or sharp scrapers, used that could scratch the surface and expose clean metal (Fig. 5.28).

Heating is always carried out by electric resistance heaters, although there is no reason why other methods should not be used, for example, gas, which is often used for this purpose in industries other than electronics. With the increasing cost of electricity in some countries alternative energy sources may well become popular in the future.

The heaters can be mounted in the pot in several ways, and these fall chiefly into three categories:

External strap on heaters, either in the form of flat strip or plate heaters, or occasionally shaped to the contours of the solder pot

Figure 5.28. Solder pots. (*a*) A total cast iron solder pot. Courtesy Electrovert Ltd.

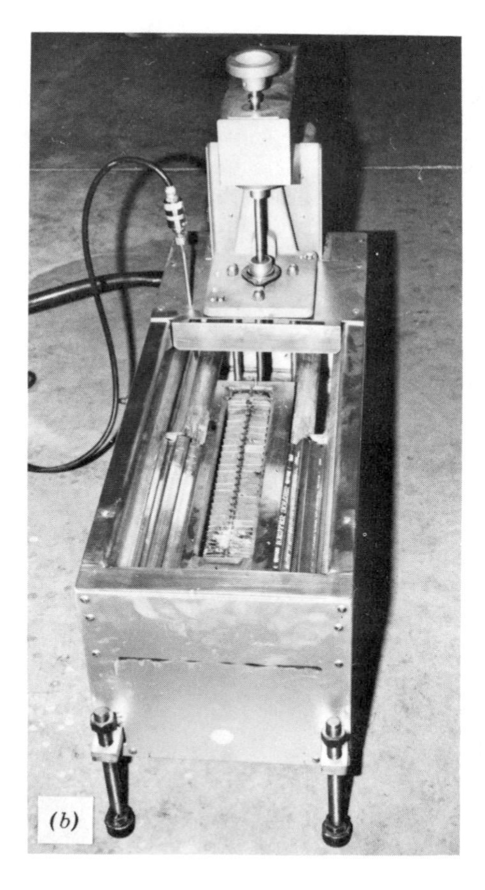

Figure 5.28. (*b*) Solder pot with a stainless steel liner. Courtesy Electrovert Ltd.

Internal heaters that are fitted into tubes welded into the pot, enabling the heater elements to be changed without draining the solder from the pot

Immersion heaters that mount directly into the solder bath and contact the solder

From the point of view of performance there is little to choose from between these different mounting methods. Efficient energy usage is much more a question of the amount and type of insulation used around the pot. Speed of heating is a function of the size of the heaters, capacity of the pot, and to a lesser extent the insulation of the pot and the venting of the machine. However, several other factors then come into the equation, such as the effect of pot capacity on the volume of solder versus the contamination levels and so on, which are discussed fully in the following pages.

Of much more importance is the question of convenience and maintenance. The external heaters do not occupy any space in the pot, and therefore the pot is somewhat easier to clean out during maintenance. The heaters that are mounted in tubes are probably the most easy to change, although if they are tight fitting and become corroded then removing the failed heater can become a problem. The most important thing is to avoid the heater system that requires the pot to be drained of solder when a faulty heater has to be replaced. Anyone who has had to use a blowtorch to melt out the solder to change a failed heater will understand the wisdom of this advice.

As mentioned, the insulation of the pot it important and must cover all the surfaces, while allowing easy access to the heaters and electrical connections. The insulation and design of the pot must be adequate to prevent the operator from accidently receiving burns through contact with the pot. The open surface of the solder should be kept as small as possible, in order to minimize the formation of dross, although there must be adequate clearance to allow for the easy removal off the dross and the accumulation of flux residues. A dross removal chute makes this much easier, and therefore more likely to be done regularly.

The solder pot is very heavy when filled with solder and must be rigidly mounted. It also requires a certain amount of vertical movement to bring the wave to the correct level. With some of the asymmetrical waves the nozzle has to be set quite precisely for a particular height of solder above the nozzle lip. Once set, only small variations in the solder wave height can be made by adjusting the solder pump. It is necessary, therefore, to be able to adjust the entire solder pot vertically. This is usually done with jacking screws at the initial installation. However, if wave heights have to be changed frequently because of variations in the design of the product, then a simpler method of moving the pot will be advantageous. This can be achieved in many ways, usually by some form of screw jacks, which are turned by a central handle or wheel. If very frequent changes to the solder pot height are required to be made, then a motorized version of the jacking system is worth the additional cost (Fig. 5.29).

Solder pot size has always been a subject of discussion, which has not lessened with the increasing cost of solder. It can cost several thousand dollars to fill a large solder pot. One special machine with a 36 in. (92 cm) solder wave required over $35,000 of solder to begin operations. Obviously if the volume of solder can be reduced without jeopardy to the process then the industry should do so. Solder volume is important for three main reasons:

Solder contamination
Thermal stability
Wave stability

Solder contamination is discussed at some length in Chapter 4. Figure 4.6 shows the way that the contamination level rises and clearly indicates that the rate of rise is a function of the volume of the solder in the pot. If the pot is very small then comparatively minor changes in the actual volume of contaminants will produce large percentage changes in the contamination level,

Figure 5.29. A mechanical solder pot jacking system. Courtesy Electrovert Ltd.

as will the addition of fresh solder. A large pot, therefore, provides a buffer against sudden changes in contamination levels. The pot size will not change the ultimate level of contamination, which is determined by the ratio of contaminants introduced to the amount of solder removed, and therefore the amount of fresh solder added.

Thermal stability of the solder wave is necessary to maintain process control. All the heat of soldering is transferred from the heating elements to the soldered joint by the liquid solder. It takes a finite time for the heaters to respond when the thermocouple calls for more heat, and during this time the only heat available is that stored in the molten solder. If the pot is too small and the volume of boards too high there could be larger temperature swings generated in the solder than would be expected from the variations seen when the pot is not actually being used to solder boards. In this respect the temperature control maintained by the controller under static conditions may not be the same as when the machine is in actual production. This is not likely to be a practical problem under normal circumstances, although the author did experience this problem on one occasion where the volume of product had been increased to such a point that the heaters were no longer able to maintain the pot temperature.

Wave stability is probably the most important factor in pot size. One manufacturer has avoided this problem by using solder jets instead of a wave in a machine with a tiny solder capacity. However, for major production, and where the advantages of the modern solder waves are desired, large volumes of solder at low pressures are needed to maintain the smooth stable wave configurations. In turn, the actual volume of solder in the wave is considerable, and the level of solder in the pot must not be lowered so much that dross is drawn from the surface of the solder into the wave when the pump is switched on. This requires somewhat more than the minimum pot size that could be acceptable for the other reasons mentioned. Therefore, it can be seen that there is no single optimum pot size. Too small can cause problems with contamination and possibly temperature control; too large and there is too much capital tied up in solder. Of course the actual solder usage will not change. Indeed, the modern soldering systems that provide such excellent joint drainage will reduce the use of solder. This cost reduction alone can more than offset the effect of the capital tied up in additional solder. Generally, if smaller volumes of product are to be processed, then a smaller solder pot size can be tolerated. The higher volumes require the larger capacity solder pots. This is reflected in the design of solder machines on the market today.

In making the evaluation of solder pot size, in fact, when making cost analyses of any of the aspects of mass soldering, always remember that the most costly part of the process is faulty soldered joints. It is false economy to try to save in any area if the result is to jeopardize quality in any way.

Automatic Solder Replenishment

Most solder wave systems can tolerate quite a wide range of solder level without causing changes to the performance of the solder wave. A variation of ½ in. (12 mm) is quite acceptable, and there should be no visible changes to the soldering quality. Usually the first consequence of a reduced solder level is that dross gets sucked down into the nozzle by the pump and produces a gritty appearance in the solder joints. Not only is this detrimental to the product, but the dross may in time clog up the pumping system or cover the heaters and contribute to their early failure.

It is useful, therefore, especially when high volumes of production are run, to have some readily visible indication of the solder level, or a warning when the minimum level is reached. It is even more useful to have some form of automatic solder replenishment. The usual method of measuring the solder level is by a float, which is arranged to operate a switch to turn on a warning light or start a mechanism to feed in more solder. It would seem that this should be a very simple thing to do considering the density of solder. Indeed the float is usually a solid piece of stainless steel that will stand exposure to the solder and not be attacked by the molten metal. However, the constant formation of oxide on the solder means that the float must be carefully designed if it is not to stick, and it must be in a separate chamber to avoid getting gummed up by the flux or flux and oil residues that are found in the soldering operation.

One elegant system uses solder wire to replenish the bath. The level sensor controls a motor that drives spring loaded grooved wheels. In turn, these feed the solder wire through a guide into the pot. Another system employs a screw jack that feeds bars down into the solder. Still another has the bar held by a solenoid, which is released when more solder is required and allows the bar to slide down to a stop under the solder surface. The solenoid locks the bar into the new position while the solder which is immersed melts and replenishes the pot. This cycle is repeated until the correct level is reached. There are obviously many variations possible, and most systems work quite efficiently.

When looking for a system to maintain the solder pot level check for the following features:

The system should be able to accept solder stock sizes of bar or wire.

The solder should be obtainable at a price that is at or very near to ingot costs; this may depend on the volume purchased.

The system should be easy to load and require little maintenance.

Any mechanism should incorporate some form of fail-safe system to avoid overflowing the solder bath.

There should be a warning when the last of the replenishing solder is used up.

The level sensor must be protected against dross and flux and oil residues and should be designed to minimize sticking.

Health and Safety Warnings

Many tests have been run to decide if there is any health hazard to the operator, or to any other personnel involved in machine soldering. There has been no indication that the mass soldering machine, correctly vented, is in any way injurious.

However, this does not mean that adequate precautions are not necessary to maintain this safety. The dross contains fine particles of solder, and the boiling action of the flux can form tiny droplets of solder when the assembly meets the wave. The following rules should be carefully observed, taught to all machine operators and maintenance personnel, and prominently displayed on all solder machines:

The solder machine must not be operated unless the venting system is in operation

Anyone looking into the machine with the doors open must wear an approved mask.

Anyone cleaning dross from the pot must wear an approved mask.

Dross must be placed immediately into an approved container that must be kept sealed.

Eating, drinking, and smoking are prohibited in the vicinity of the solder machine.

Any persons handling solder or soldered parts must wash their hands before eating, drinking, or smoking.

Work closely with your safety officer and state and federal authorities. The above are basic proposals and the absolute minimum precautions to be taken. Because each installation is different, safety measures should be checked with the authorities on the spot.

Machine soldering is a very safe operation. There is much more possibility of harm through ingesting lead through the skin than by breathing the air near a properly vented soldering machine. For this reason, anyone handling solder or soldered items must pay particular attention to their personal hygiene and, if necessary, wear gloves.

While the above has been primarily concerned with the dangers of lead poisoning, do not forget the other possible dangers to health. The solder pot is hot. Splashes of solder are painful and can cause serious burns. Always wear safety glasses when working near the solder pot, or preferably a full face mask. Clothing should cover as much of the body, arms, and legs as possible and must not be made from synthetic materials which can melt and adhere to the skin if subjected to the heat of molten solder. Never allow water near the molten solder; it will cause an explosion if it contacts the surface, and molten solder will be blasted everywhere. Take care to see that any mechanical moving parts are protected against accidental contact.

A secondary source of danger is electric shock from damaged or leaking insulation in the machine. A good electrical ground, checked at regular intervals, will prevent this as well as any possible damage to components or circuitry in the product being soldered, from stray electrical currents.

The Solder Pump

In order to produce the wave, the solder has to be pumped through the nozzle. This is almost always done with a simple centrifugal pump, which is usually driven by an electric motor. The pump has to operate in a hostile environment, being subject to the hot molten solder and any flux or flux residues that may find their way into the pumping area. The operating temperature alone makes it almost impossible to put bearings into the solder, and therefore all the designs arrange for the bearings to be well above the solder surface. Even here the working temperatures prevent the use of the traditional lubricated surfaces, and carbon or other high-temperature bearings are used nearest to the solder surface.

Even these must be suitably protected from the dross or other residues that float on the surface of the solder. These residues will certainly cause damage if they are allowed to get into the bearing surfaces (Fig. 5.30). Similarly, the solder level must not be allowed to get so low that the dross or other residues that collect on the surface of the solder are dragged down into the pumping area. They can be forced into the nozzle, where they become deposited and cause clogging of the passageways and a subsequent deterioration of the wave. They will also get into the soldered joints and can cause a

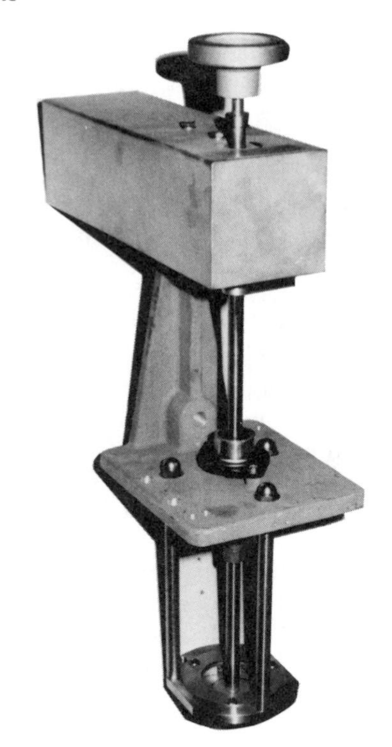

Figure 5.30. Two typical centrifugal solder pumps. Courtesy Electrovert Ltd.

gritty appearance, which is sometimes confused with that caused by metallic contamination of the solder.

The actual wave height is usually controlled in one of two ways: either the rotor of the pump is moved in relation to the static housing in which it is mounted, or the rotor is fixed and the speed of the motor is varied. Occasionally both methods are used, mechanically moving the rotor to make major adjustments and using the electrical speed control to make the final setting. The motor may be a standard AC single phase 1/8 to 1/2 horsepower, or where speed control is a function off motor speed, a DC motor and speed controller is usually used. In a few cases the wave is adjusted by some form of valve or flap in the solder passage from the pump to the nozzle, which diverts a portion of the solder directly back to the pot or restricts the volume of solder that reaches the nozzle by reducing the area of the entrance to the pump or to the nozzle itself.

A few manufacturers use an air motor to drive the solder pump and claim that it has advantages in smoothness of operation and speed control. There is in practice no advantage in any particular method driving the pump, as long as the speed, once set, is constant, the setting can be repeated accurately, and the wave height can be controlled easily. It is, of course, absolutely essential that the wave itself should not have any short-term variations in height or shape, for any cause, and any long-term variations must be small, predictable, and easily corrected by cleaning, adjustment, or other simple maintenance operations.

Although in the preceding paragraph the control of wave height is mentioned, it must be emphasized that this refers primarily to the older style of symmetrical or parabolic wave. With these wave shapes the change in height affected only the depth of the wave available for soldering and the amount of dross generated. The higher the wave, the greater the turbulence in the pot, and the more dross. The actual shape of the wave stayed the same. With the newer asymmetrical waves, the rate of flow will change the wave shape and the flow patterns, from which the wave derives its superior performance. The wave height, therefore, must be set to produce the required wave shape and the relationship between the wave surface and the board to be soldered adjusted by jacking the entire pot. Minor changes can be made by adjusting the wave itself, but only if this does not affect the shape or flow of the solder.

One manufacturer has produced a pumping system that utilizes the magnetic forces produced in an electrical conductor when a current is passed through it. The solder passes through a powerful electromagnet, which induces a current into the solder stream. The inductive reaction between these currents and the magnetic forces of the electromagnet propels the solder through small jet nozzles to produce a series of solder streams, or what is called a hollow wave. This form of pumping does not have any moving parts and, of course, avoids any problems of contamination or lubrication of shafts or bearings.

This particular machine can also be used with an oil intermix in the wave, with oil injected from an internal reservoir. The solder capacity is quite small, and the entire contents of the pot are circulated about 10 times per minute. It is claimed that the high speed of the solder in the hollow wave produces a mechanical scrubbing action on the joint, which assists in cleaning off any contaminants and assures total wetting. It is also claimed by the manufacturer that the solder jets possess a large degree of flexibility, and thus will conform to the surface of any twisted or warped boards. Of course, with the small pumping power available with this system, it is not possible to pump large volumes of solder, and thus the normal wave shapes are outside the capability of this form of solder pump.

No matter how well the pumping system is designed there will always be a need from time to time to clean out the pump and pump chamber. The ability to remove the pump easily, without draining the pot, is essential. With solder pumps, simplicity is always best. Maintenance of the pump and drive should always follow the manufacturer's recommendations and must be carried out regularly. Motors and pulleys should be checked for alignment to prevent abnormal belt wear. The solder pump is normally one of the least troublesome parts of the solder machine, and if correctly set up and maintained will be totally trouble free.

The Solder Wave

The solder wave is the heart of the solder machine. The moment that the board touches the solder the formation of the solder joint begins. If the design of the nozzle is not correct for the assembly to be soldered, or is not correctly set up, then all the preceding operations will have been for nothing and there will be no way to achieve the objective of *zero defect soldering*.

When wave soldering was originally developed, it was a tremendous advance in productivity over the traditional hand soldering and the then popular manual dipping of printed wiring assemblies. Certainly, there were occasional short circuits, and icicles were very evident, but with such an improvement in output there were few complaints, and these problems were largely accepted as part of the process. The wave of those days had a simple symmetrical shape. It was generated by pumping solder through a nozzle that took the form of a simple rectangular slot projecting a short distance above the surface of the solder bath. The solder fell back into the bath in a natural parabolic curve, and no attempt was made to shape the wave in any way (Fig. 5.31).

As the industry grew, wave soldering became more widespread as a manufacturing process, and serious attempts were made to reduce the incidence of soldering problems, as well as to increase the speed at which soldering could be carried out, without jeopardizing the quality of the soldered joints. This resulted in the development of many different wave shapes, and other ideas for controlling the flow of the solder during the actual soldering opera-

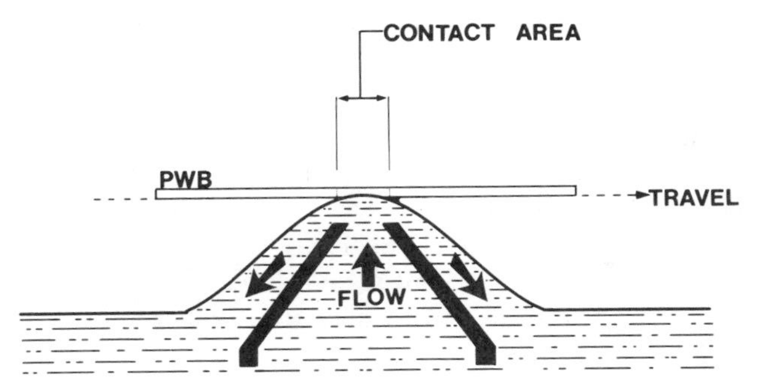

Figure 5.31. A diagram of a symmetrical parabolic wave. Note the very small contact length between board and solder.

tion. Many of these fell by the wayside, but several were developed further and are the basis of the various wave designs available today.

Without doubt, the two developments that have had the most impact on the industry are the use of oil in the wave and the invention of the asymmetrical adustable wave. The defects induced by the original symmetrical wave shape were not of major importance with the old designs of PWB. These were typically single sided with large component mounting pads and wide spacing of conductors. However, packaging density increased, and plated through holes and close conductor spacing became the order of the day. Fluxes and solders improved in performance, and the deficiencies of the solder machine became more apparent. The increasing use of electronics in the space program and in mililary and aviation products brought the importance of solder joint reliability to the fore. It was recognized that the solder wave contributed to two soldering problems.

Excess solder, causing short circuits between adjacent conductors, and solder spikes or icicles.

Inadequate solder, resulting in a lack of a top side fillet in plated through hole boards, and unwetted or "dry" joints. These problems could be eliminated or reduced only by running the soldering machine at a very low speed and were obviously a result of the inability of the wave to heat the joint.

These faults occurred when the joints were otherwise totally solderable and were obviously tied up with the inability of the wave to drain off the excess solder and, as already mentioned, to bring the joint to soldering temperature quickly.

The solder machine designers tackled these problems in several ways:

Using oil in the wave to reduce the surface tension and permit the solder to drain more easily

Increasing the contact length of the solder with the wave so that there was more time for the heating to take place

Using the natural forces of gravity and surface tension to assist the excess solder to drain from the board

The use of these methods will become apparent in most of the different types of nozzles that are to be discussed. This is not meant to be a complete listing of all the nozzle types available, but rather a description of the design and operating characteristics of the chief types of nozzles found in today's commercially available machines.

The Symmetrical Wave The symmetrical wave was the original solder wave, and is still occasionally used today. However, it has been improved to produce better soldering results than those of the early days. The first action was to widen the wave, thus producing a longer contact with the board. The second was to use an angled conveyor to reduce the incidence of shorts and icicles. Since this is an important factor in many of today's soldering machines it will be examined in detail, especially as the basic concepts involved relate also to many of the other wave shapes.

It was soon noticed that the formation of shorts and icicles, in fact, all of the ills connected with excess solder, could be related to the way in which the board exited from the wave. With the horizontal conveyor and a symmetrical wave, the board leaves the solder in a region where it is falling rapidly back into the solder bath. It is thus moving away from the board at a high speed and a considerable angle. This is conducive to the solder on the joint cooling quickly. The connection between solder in the wave and the solder in the joint is broken rapidly, and there is not time for the solder in the joint to drain back naturally.

As anyone who has used a hand soldering iron will understand, it takes a finite time to drain the excess solder from a joint. In manual soldering the iron is held under the joint, and once the solder is molten the iron is slowly withdrawn allowing the solder to drain down onto the bit. The fact that the excess solder will hang as a drop on the soldering iron shows that the surface tension is sufficient to prevent it falling off by gravity alone (Fig. 5.32). Therefore, the force of the surface tension must either be reduced, as with the use of oil in the wave, or utilized to assist solder drainage by the correct design of the wave to board interface.

At the center of the wave, because the solder is falling on either side, there exists a small area where the motion of the solder is zero and the surface of the wave is almost flat (Fig. 5.33). By moving the conveyor from the horizontal to a suitable angle, the board can be made to exit the wave in this spot. The conditions are now quite different from those that exist with the horizontal conveyor. The joint emerges from the solder at a small angle from the solder surface. The solder is almost stationary relative to the board movement, and the surface tension of the wave will tend to promote the

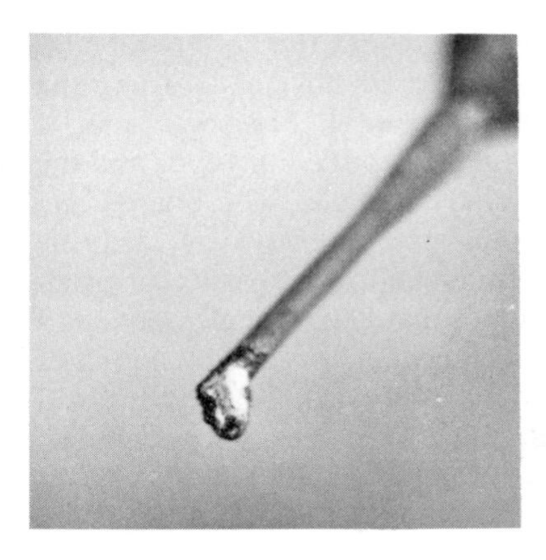

Figure 5.32. The strength of the solder surface tension holds this heavy drop of molten metal on the soldering iron bit.

drainage of the excess solder from the joint. It was found that this arrangement markedly reduced the incidence of shorts and icicles, and much of the following design work on nozzles has been based on this fundamental concept.

The use of the angled conveyor is not without some penalties. The board now meets the wave at a point where the solder is flowing fast, and any maladjustment in board height or a warped board can result in the board diving into the solder and flooding the upper surface. This almost always results in a scrapped assembly. The conveyor also exits the machine at an inconvenient height. However, the use of this principle of controlling the exit angle of the board from the wave has so many advantages that until

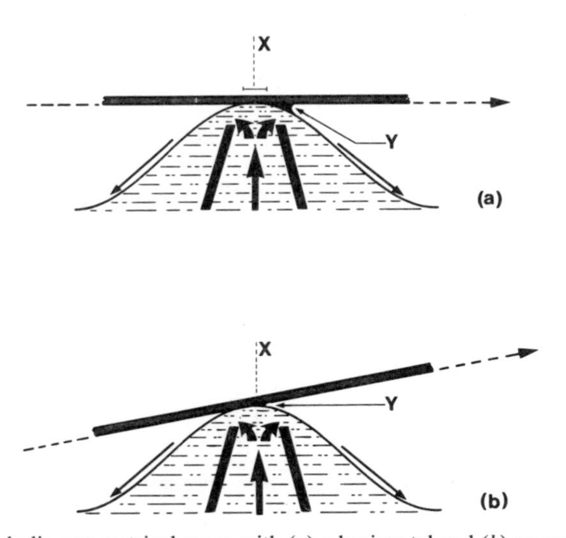

Figure 5.33. A parabolic symmetrical wave with (*a*) a horizontal and (*b*) an angled conveyor.

some new development offers the same advantages with the horizontal conveyor, most soldering machines will continue to be designed to use the conveyor set at some small angle from the horizontal. For some special purposes the additional solder produced by the horizontal conveyor and the symmetrical wave is beneficial, for example, with single-sided boards and large hole diameters. Some machines have an adjustable conveyor, as it was believed that this should be set to an optimum angle for the particular product being processed. More recent work indicates that a fixed angle between 4 and 6° is universally acceptable, with the smaller angle producing a greater contact width between board and solder when used with the asymmetrical adjustable wave. This provides faster soldering and better solder drainage, eliminating solder shorts and other forms of excessive solder.

The Wide Symmetrical Wave With the understanding of the wave characteristics already discussed, there were obviously advantages in making the wave as wide as possible, by opening the nozzle as far as was practical. With the angled conveyor this offered a larger horizontal section of the wave for the board to exit, although there was still only a small area of stationary solder. However, with a horizontal conveyor the contact area between the solder and the board was markedly increased, permitting faster soldering (Fig. 5.34). Where simple boards are to be soldered at high volume, this latter system has certain advantages and should be considered as a viable process.

In order to produce a wider wave, the nozzle opening could be made wider, but there is a limit to this width, depending on the solder wave height desired and the amount of pumping force available. Therefore, the use of support or extender plates became common. These supported the solder after it emerged from the nozzle and produced a wider wave without pumping a greater volume of solder.

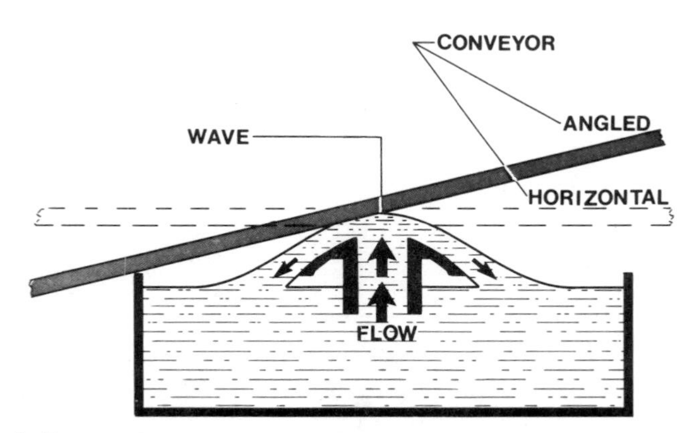

Figure 5.34. A diagram of a wide symmetrical wave. Note the greater contact length between the board and the solder.

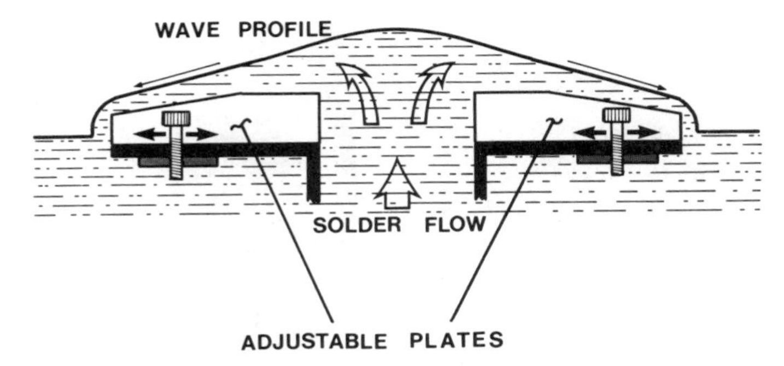

Figure 5.35. A diagram of a typical adjustable symmetrical wide wave.

The Adjustable Symmetrical Wave Once the use of extender plates became popular, it was only a matter of time before the idea of making them adjustable was developed. This provides a wave that can be varied in width and adjusted for the optimum for any particular purpose. When used with a horizontal conveyor it can be set to produce almost any contact length with the board. It offers very high speed soldering for simple boards, for example, single-sided with straight through pins such as connectors. It is also very effective for tinning the leads of components loaded into pallets. As the extender plates are separated more widely the shape of the wave changes, and the center ''hump'' can provide a useful wave shape for soldering boards and components with longer leads. The long entrance section of the wave provides a certain amount of preheat and may provide the solution to soldering boards with a high thermal mass (Fig. 5.35).

Figure 5.36. An asymmetrical supported wave. Courtesy Light Soldering Development Ltd.

Figure 5.37. The nozzle and pump used to produce the solder wave shown in Fig. 5.36. Note the plate to support the solder wave. Courtesy Light Solder Development Ltd.

The Asymmetrical Wave By using an extender plate on one side of the nozzle only, the wave could be shaped to provide improved characteristics for both the exit section and the contact area. This "supported, asymmetrical wave" is well illustrated by Fig. 5.36, which shows it used with a horizontal conveyor. The nozzle used to produce this wave is shown in Fig. 5.37, where the wide nozzle opening and the wave supporting surface are clearly shown. Another form of this wave is shown diagrammatically in Fig. 5.38. There are very many different forms of this wave shape, used with

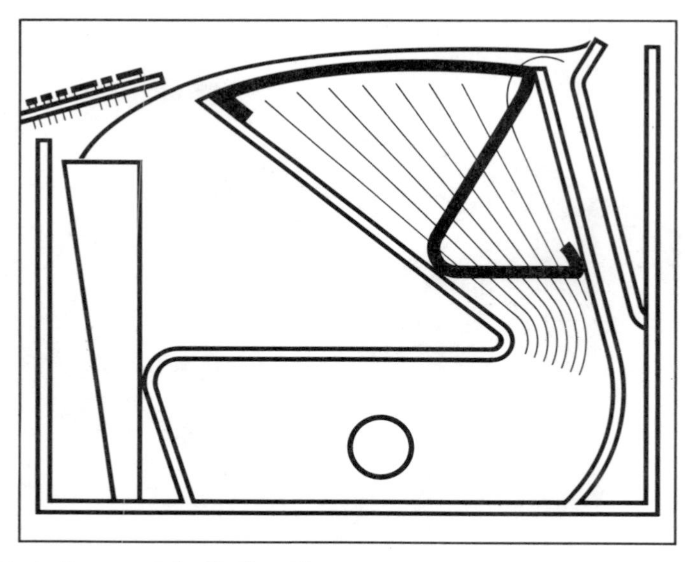

Figure 5.38. A diagram of the Hollis "Z" wave. Another development of the asymmetrical wave. Courtesy Hollis Engineering.

both horizontal and angled conveyors. It is quite impossible to describe all these variations; some are produced by supporting the wave, some by shaping the nozzle, and most use both principles to achieve their results. The only way to assess the advantages and disadvantages for any particular purpose is to carry out trials on the product to be soldered.

The Adjustable Asymmetrical Wave The adjustable asymmetrical wave is the ultimate result of the years of steady development in nozzle design. As discussed in the earlier description of the symmetrical nozzle, there are two functions in the effective soldering of a PWB. First, it is necessary to heat the joint quickly and wet it with solder. Second, it is necessary to remove all surplus solder. In this wave, for the first time, these functions are separated and carried out by individual portions of the wave shape. Each portion is designed specifically for the particular task that it has to perform (Fig. 5.39).

This wave shape is always used with an angled conveyor set between 4 and 7°. In the nozzle the solder flow is arranged so that the majority of the molten metal is directed toward the oncoming PWB. The small volume of solder flowing in the opposite direction is used to fill an almost static lake, which overflows slowly on the opposite side of the nozzle to the main flow.

As the board approaches the solder it is first met by the rapidly flowing main stream. This quickly heats the joint and imparts some washing action to the surfaces to be soldered. As the joint components reach the melting temperature of the solder, it will wick up into the joint, wetting all the surfaces, and form a sound electrical and mechanical connection. By this time, the board will be leaving the fast flowing main stream and enters the almost stationary solder lake. Like all free static liquids, the surface of this lake is horizontal, and because of the angle of the conveyor the board will leave this surface at a very small upward tilt. This small change from the horizontal is sufficient to allow the solder to drain from the joints because it

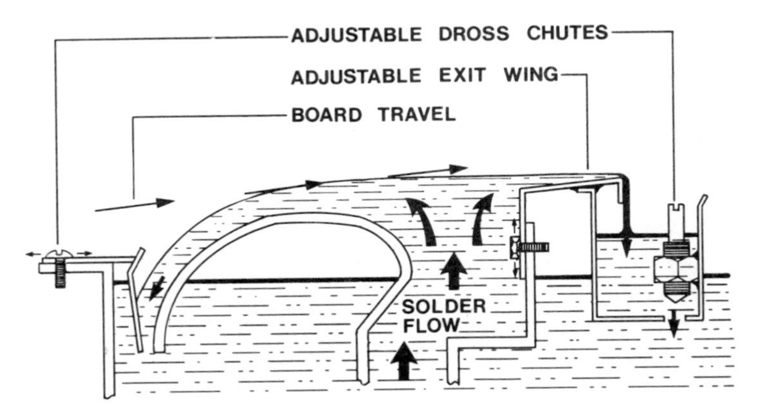

Figure 5.39. A diagram of the adjustable symmetrical wave.

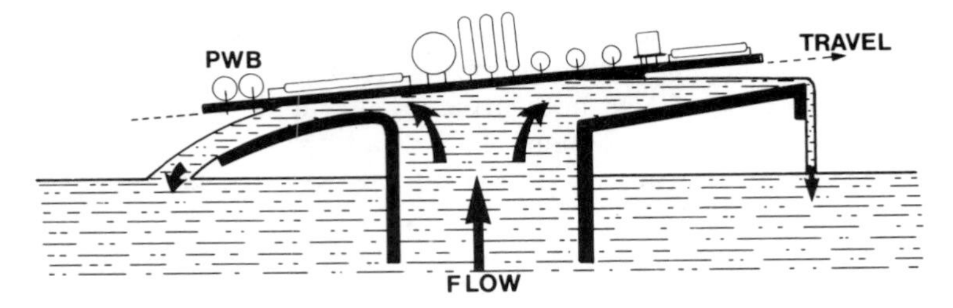

Figure 5.40. A diagram of a printed wiring board passing over an adjustable asymmetrical wave. Courtesy Electrovert Ltd.

provides a very slow vertical separation from the solder surface. The considerable surface tension of the solder, which is applied to the joint as long as there is contact between it and the surface, materially assists in this removal of the excess metal. Some solder remains because of the opposing forces inherent in the actual wetting of the joint, as discussed in Chapter 2 (Fig. 5.40).

The wave, therefore, solders quickly and drains the excess solder effectively. It is one of the most useful general purpose waves, is very effective for the tightly packaged boards in modern electronics, and with solderable components and the correct processing parameters, is capable of the highest levels of quality and freedom from soldering defects.

This wave is not adjustable in height by changing the pump speed and requires setting of the nozzle adjustments for any particular wave height required. Once set, only very minor changes to the wave can be made with the pump setting, since these will change the wave configuration. The main height adjustment is made by jacking the entire solder pot. The maximum wave height is approximately 3/4 in. (10 mm) although there are special deep wave versions available. These wave types require careful adjustment to take full advantage of their maximum capability, although once set they do not require frequent readjustment.

The adjustment of the back plate, or adjustable exit plate as it is sometimes called, must be made very precisely. First set the wave height so that the longest leads to be soldered will clear the nozzle. Then adjust the back plate until the following conditions of solder flow are obtained.

1. When the wave is switched on there should be no solder flowing over the back plate.
2. If a board is run over the wave, or if the solder is pushed very gently over the back plate with a spatula, the solder should flow over the plate in a very thin stream and continue to flow until the solder pump is switched off or the wave height lowered.

This setting is extremely precise and must be carried out every time the wave height is changed. It can be seen therefore that once set up correctly the wave height must never be haphazardly readjusted or "fine tuned." An incorrect setting will reduce the ability of the wave to remove excess solder and increase the possibility of solder shorts or bridges.

The Dual Wave and Soldering SMDs Much has been written concerning the use of Surface Mounted Devices (SMDs) that is exagerrated or even incorrect. Therefore before discussing the use of the solder machine for attaching these components it is necessary to bring the subject more clearly into focus.

Surface mounting is not a new technology. It has been used in one form or another for many decades, ever since the flat pack gained popularity. It is not a single technology but a collection of many different technologies. These range from the complex assembly of multileaded components mounted on both sides of a board using reflow techniques, to the simple use of discrete chip components mounted on the solder side of the board and soldered at the same time as the leaded parts. Because this book is primarily concerned with the use of the solder machine to make reliable joints the latter form of construction will be the SMD technology discussed in this section.

In this assembly method the SMDs are attached to the solder side of the assembly using an adhesive, usually a fast curing epoxy. The entire assembly of both leaded and SMD components is then soldered in the normal way as if it contained leaded parts only. Of course the SMDs will pass completely through the solder wave, but if correctly processed excellent reliable joints will be formed between the SMDs and their mounting pads. The leaded components will of course be soldered in the usual way.

This is the simplest and lowest-cost method of using SMDs. With care it is possible to solder boards of this type with no solder defects. However there are certain factors which must be considered. First the component packaging is usually extremely tight, with minimal conductor and component spacing. The solder wave must therefore be able to remove all excess solder to eliminate the possibility of solder shorts. This becomes a major problem when attempting to solder multileaded components and may require the use of an air knife system (Page 116).

Most of the modern solder machines will provide bridge-free soldering of the simple chip components. Another problem however is found when using the standard wave systems. Some joints will be found that have not wetted at all; in fact it will be obvious that the solder has not even touched the joints. This defect is called *skipping*.

In the early days of soldering SMDs this was thought to be caused by outgassing of the flux, which formed a bubble around the joint and effectively prevented the solder from contacting the parts to be soldered. The use of very low solids fluxes was proposed, which of course made bridge-free soldering more difficult. Finally it was discovered that the answer was very simple, and related to the flow pattern of the solder wave. For many years the main effort had been to design solder waves that were capable of removing excess solder so that the increasingly tightly packaged assemblies could be soldered bridge free. This involved making the surface of the waves as smooth as possible with a mirrorlike laminar flow. As a joint passed through this form of wave it had a shadowing effect, effectively making the solder flow around a following item that was very close to the board surface. The problem was not apparent with leaded joints as they protruded sufficiently to cause local turbulence which prevented complete shadowing. With SMDs however the problem became acute with a large percentage of joints on any assembly being totally free of solder. It was discovered that making the solder surface turbulent mitigated the problem, but then the ability to solder bridge free was reduced.

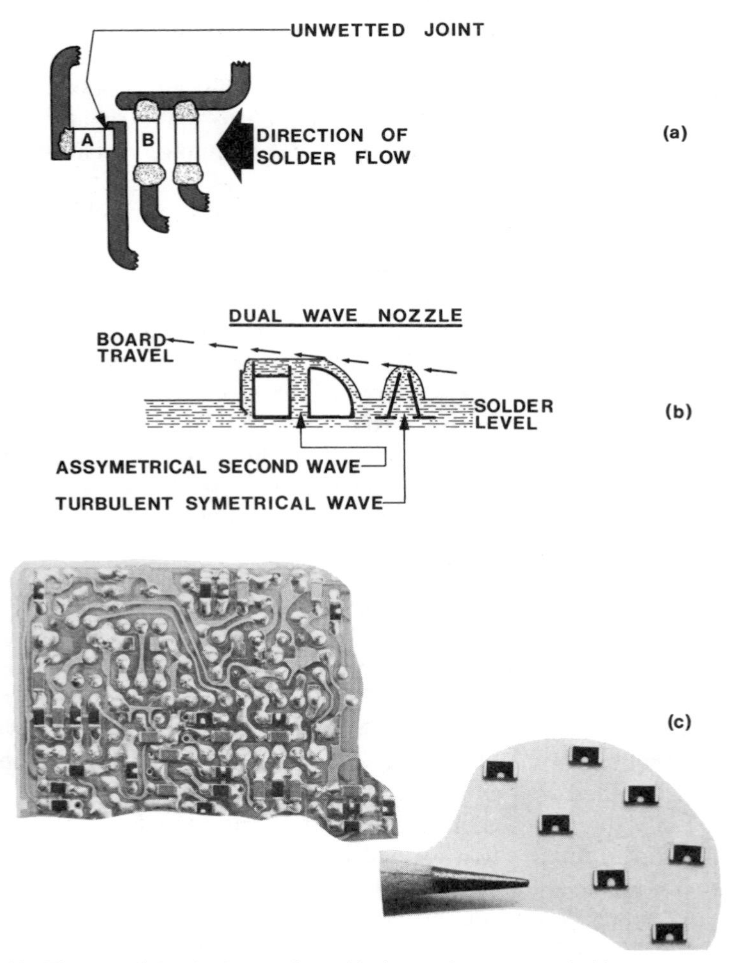

Figure 5.41. The use of the dual wave for soldering surface mounted chip components. Courtesy Electrovert Ltd.

Figure 5.42. (*a*) A typical dual wave system. Courtesy Electrovert Ltd. (*b*) A photograph of the nozzle that produced the wave shown in Fig. 5.42*a*. Courtesy Electrovert Ltd.

The ultimate solution took several forms. The most popular was to provide two waves, the first very turbulent to wash the joint in all directions and provide adequate solder coverage. The second wave was smooth and removed the excess solder, so preventing solder shorts or bridges (Fig. 5.41). In some cases a dual nozzle assembly and one pump produce these effects, but in most cases a dual nozzle dual pump arrangement is used (Fig. 5.42). Any form of turbulence will get rid of skipping and in some cases the single wave is made turbulent at the leading edge. One system bubbles nitrogen gas through the wave. Another vibrates the leading edge of the wave with an electrically driven oscillating blade (Fig. 5.43).

A final problem that can occur in soldering SMDs is called *leaching*. If the component is left in the solder for any length of time the molten tin in the solder will dissolve some of the metallizing that provides the connection on the device. A typical specification may indicate that 30% of the connection will be lost if the part is in the solder for 10 seconds. The solution is comparatively simple: to solder quickly and limit the exposure to the solder to 5 seconds or less. In practice this means preheating to a temperature on the top laminate of not less than 220°F (94°C) and soldering at a speed which limits contact time with the wave. Under these circumstances leaching is

Figure 5.43. This picture shows the vibratory pattern generated in the wave to eliminate solder skipping when soldering surface mounted devices. Courtesy Electrovert Ltd.

seldom a problem; however any hand soldering must recognize similar limitations.

Soldering SMDs on the solder machine is not difficult, but does require strict process control, perfect solderability, and careful handling. Although some hand rework is possible it is difficult and frequently results in damage to the product. If a Zero Defect Soldering program is not in operation it should be set up before any form of SMD work is contemplated. Any surface mounted assembly must be looked upon as being largely noninspectable and nonrepairable.

Dross and Dross Chutes Dross is produced whenever the surface of the molten solder is exposed to the atmosphere. When the surface is agitated as in the solder wave, the rate of dross production is increased. However, most of the dross is generated in the turbulence formed by the solder stirring up the surface of the pot as it returns back from the wave. To reduce this effect channels or chutes are incorporated into some nozzles (Fig. 5.44) to reduce this turbulence by funneling the solder down under the surface, and not allowing it to stir up the main body of molten metal. These dross reduction chutes are quite effective, but they must be kept clean to permit the proper

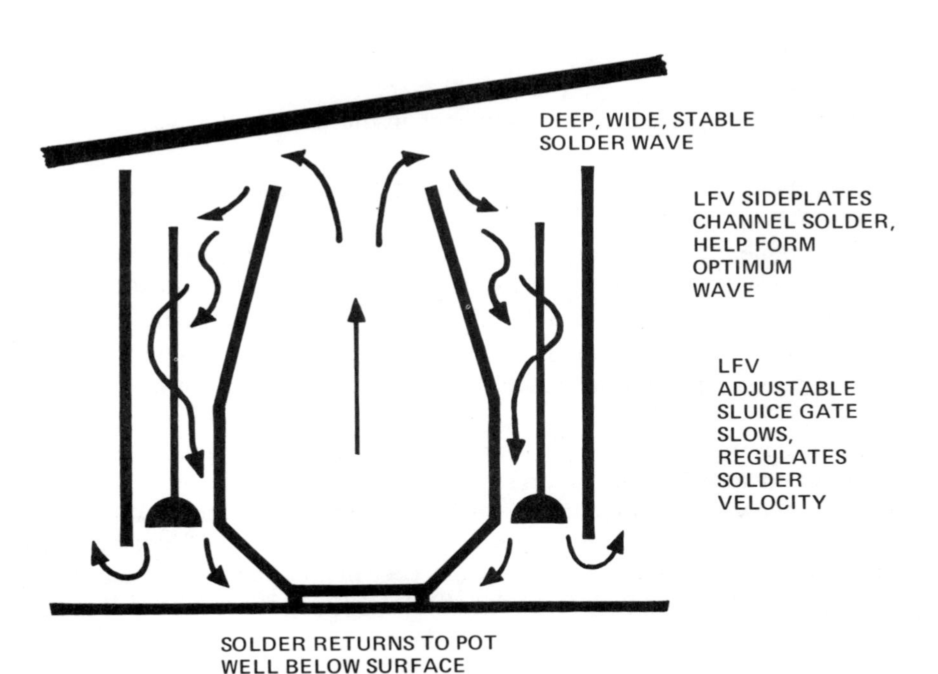

Figure 5.44. A typical dross reduction system. Courtesy Hollis Engineering.

flow of solder. The surface of the solder pot can be covered with a layer of high-temperature oil to reduce the formation of dross by excluding the air. This of course occurs automatically when oil is used in the solder wave. Glass beads have been used as a "cover" on the solder, and indeed anything that floats on the surface, can withstand the temperature, and is inert to the solder can be used to retard the formation of the oxide that forms into dross. Generally, it is found that the mess and bother of using these cover layers outweigh any savings that accrue.

Dross is usually seen in the solder pot as a silvery slushy material, which can be scraped together into large solid lumps. If these lumps are left alone they become covered with a black powder. In fact, what is commonly called dross is chiefly solder. As the molten metal falls onto the solder surface it breaks up the thin oxide covering, which is tenacious enough to cover the small solder particles formed by the solder splashing and other turbulence. These oxide covered balls of solder stay on the surface and collect together to form the dross as it is commonly called. Together with flux and flux residues, after some time of running the solder pot becomes covered with this thick rather solid layer. This is why dross chutes work so well, since by channeling the flow of solder below the surface the turbulence and splashing is reduced. This also explains why, in general, the higher the wave, the more dross is generated.

There are several factors that affect dross formation:

Increasing the solder temperature will generally increase the rate of dross formation.

Anything that increases the turbulence in the wave will increase the formation of dross.

Impurities in the solder can contribute to an increase in the dross formation rate.

Too frequent skimming of the solder pot is a sure way to increase dross formation. A thin layer of oxide will help to prevent further oxidation.

Too low a level of solder can sometimes induce turbulence into the solder surface in the pump area which will increase the formation of dross.

There are proprietary chemicals available from most solder suppliers, which, when sprinkled on the dross, reduce the surface tension of the oxide films and allow the majority of the solder contained in the dross to fall back into the pot. The large mushy deposits turn into a thin layer of black powder which is easily removed. Weigh up the cost of the chemicals against the savings in solder, not forgetting the cost of labor, and decide if this is a profitable way to go.

Remember that the untreated dross is largely solder and therefore has a market value. Your solder supplier will usually take back the dross at a standard price per pound. Dross does contain lead in the form of small particles and dust, and therefore the safety rules already discussed must be adhered to, especially when dross is handled in any way.

Selecting the Right Nozzle As discussed, there are many variations in the design of the various wave shapes and the nozzles that produce them. Without some plan the selection of the correct nozzle can become a bewildering problem. There are so many variations in board design, component mounting, production volume, and other parameters that what may be suitable for one set of conditions may not be the best choice for another. The only real way to make the decision is by testing the process under actual operating conditions. Any vendor worth doing business with will only be too pleased to arrange for the testing to be carried out on the actual product to be soldered. However, the testing must be carried out in a logical manner, and the following items are suggested as the basis of a test checklist.

Make tests on the most difficult board to be soldered (this is usually the most densely packaged).

Check that the nozzle can be interchanged with other types, since future product changes may require other solder wave shapes.

Be sure that the nozzle can be changed without having to drain the pot.

Look for dross reduction chutes and check the rate of dross production.

Look for a smooth wave without ripples; remember that some waves are deliberately turbulent for specific work.

Check for wave stability. There should be no changes in wave height or configuration even with changes in the level of solder in the pot. A minimum of ½ in. (12 mm) solder level change is acceptable.

Make sure that the nozzle chosen is compatible with the available conveyor angle.

It is not usually possible to select a nozzle that will do everything required; compromise on some of the parameters may be necessary, but excellence in soldering must be the prime concern and other considerations must always be secondary.

Oil in the Solder Wave

As mentioned previously, one of the earliest methods of reducing the incidence of solder shorts was to use oil in the solder wave in order to reduce the surface tension. This is still in use, especially with the symmetrical or bidirectional wave. In addition to assisting the excess solder to drain from the joint, the oil eventually flows onto the surface of the solder pot and assists in reducing the amount of surface oxidation of the solder. Oil can be introduced as a thin surface layer on the wave, or intermixed with the solder at the impeller of the pump, to obtain a uniform dispersion of the oil in the solder wave (Fig. 5.45).

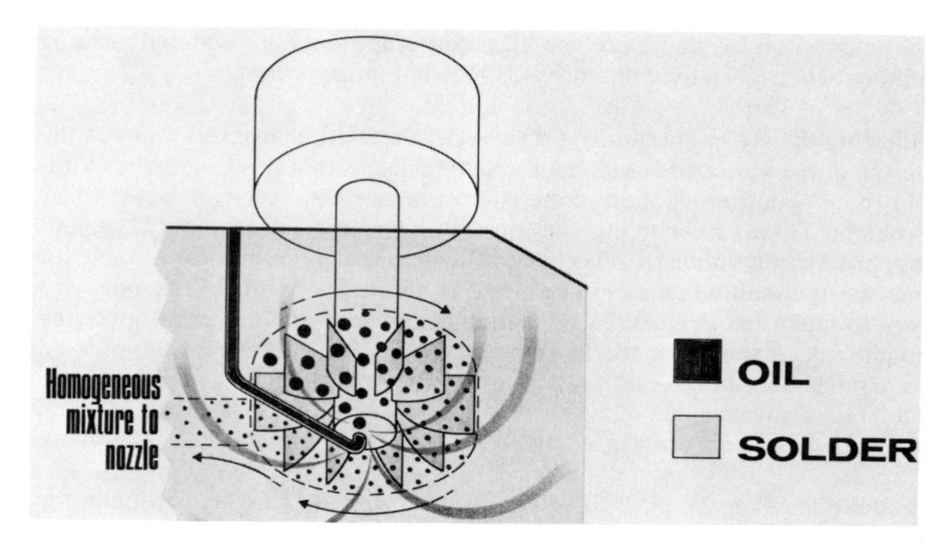

Figure 5.45. A diagram of a typical oil injection system. Courtesy Hollis Engineering.

It is also claimed that the oil speeds the wetting of the solder to the metallic parts of the joint, and thus speeds up the soldering process. This wetting should not be confused with the solder ''wetting,'' which forms the actual joint and is of course dependent on the action of the flux in removing the oxide from the metals to be joined. The point is also made that the oil immediately coats the newly formed joint and retains the bright shiny surface. In addition, it is claimed that the oil mixes with the rosin in some fluxes and makes their removal easier.

There is no doubt that oil in the solder wave does materially remove many of the shortcomings of the symmetrical wave and improves the performance of the asymmetrical waves that are designed to be used with oil. The oil is usually introduced into the solder wave in one of two methods. In the first, it is pumped or allowed to flow by gravity to a nozzle located in the pump or pressure side of the wave. The oil forms small bubbles or globules, which burst on the wave surface and form a thin layer over the solder. The oil then flows out onto the wave surface of the solder bath and eventually overflows into a suitable container. This is a total loss system, and the oil is only used once. With modern high temperature oil there is no reason why they should not be reused several times.

This is done in the second and more usual method. In this system the oil from the wave is taken back from the surface of the solder through a special valve, which controls the flow back to the region of the pump impeller. Here, the oil and solder are thoroughly intermixed and together returned to the wave. The oil again forms a thin layer on the wave surface, and together with the small particles of oil mixed in with the solder effectively reduces the surface tension and assists in the wetting of the joint and effective drainage of the excess solder (Fig. 5.46).

When the wave is shut off the level of solder in the pot rises, and the oil on the surface runs into an overflow compartment from which it is drained into a container for disposal. Fresh oil is then added when the wave is once again started. With the correct tinning oil, one fill of about 1 gal. (4.55 liters) will last for a full 8 hour shift.

As with all soldering systems the efficiency of this process will depend on the correct setting and maintenance of the equipment, and the manufactur-

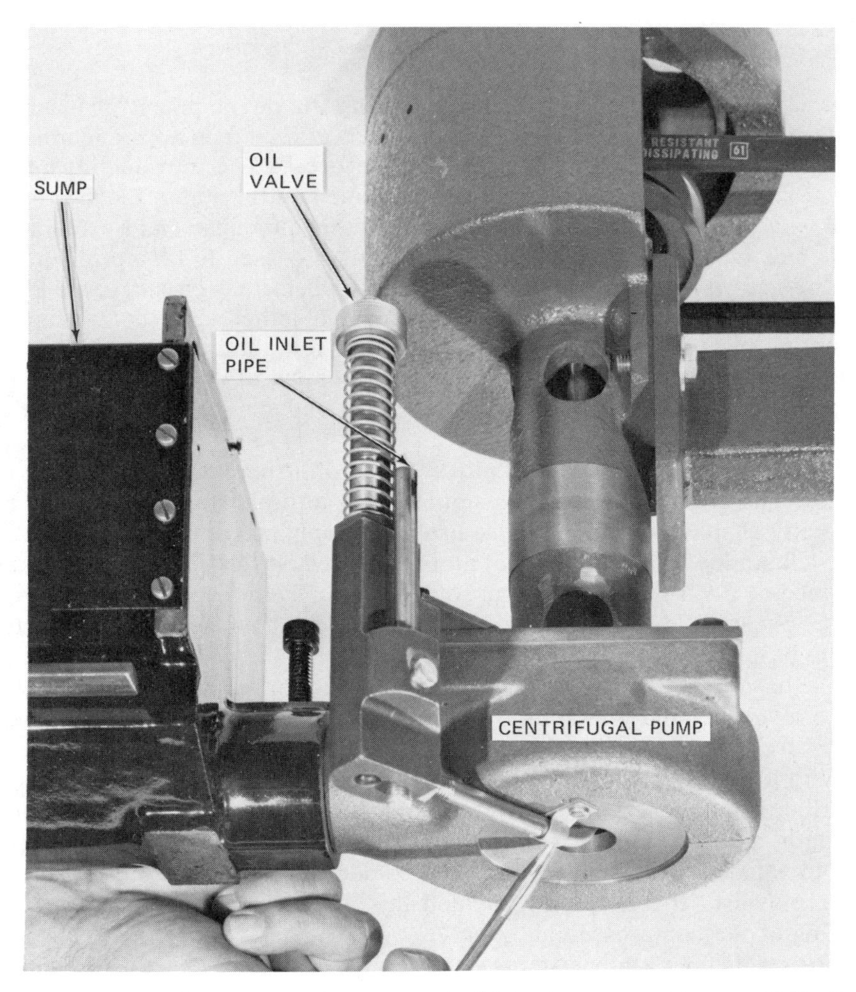

Figure 5.46. This picture shows the pump and oil injection system. Courtesy Hollis Engineering.

er's instructions must always be followed carefully. Once the decision to use oil injection in the wave has been made, the oil becomes a critical part of the process, and the control of the oil system and the correct addition of the oil must be carried out with the same thoroughness as the flux is controlled.

At one time, there was some concern raised that the oil in the wave would result in oil becoming imprisoned in the solder in the joint. If a joint is crosssectioned inclusions in the solder will be found no matter what form of soldering has been used. If oil was used in the wave these inclusions will contain oil. However, tests have shown that this oil has no effect on the joint either mechanically or electrically and will not reduce the life or reliability of the product.

The Air Knife in Soldering

As was discussed in earlier chapters, during the development of machine soldering, the presence of solder bridges or shorts between adjacent conductors was one of the major defects of the process. The introduction of oil injection into the wave and later the development of the adjustable asymmetrical wave were both intended to overcome this difficulty, and by and large this objective was achieved, and solder bridges ceased to be a problem.

However, with the ever tighter packaging of electronic circuitry, the PWB conductors are becoming smaller and closer together, and the spacing of mounting pads and terminals is reduced to a few thousandths of an inch. Solder bridging is, therefore, once again becoming of concern, and the air knife is being offered as another means of reducing or eliminating this problem.

In its simplest form the air knife consists of a tube with a slot along its length. The edges or lips of this slot are only a tiny distance apart, and by correctly shaping and sizing these lips and supplying compressed air to the tube a fast moving fan or sheet of air is produced. If this sheet of moving air is aimed at PWB that still has the solder molten from the wave, it will wipe off the excess solder, leaving only that which is held in place by being wetted to the base metal (Fig. 5.47).

The basic process is not new. It has been used for many years with heated air to level the solder coating on bare PWBs and for removing the excess solder from wire or strip copper during the tinning process.

With loaded PWBs, however, the task is not as simple. A certain amount of solder has to be left on the joint to provide the fillet that gives mechanical strength, and the variability of the joint structure makes it much more difficult to achieve consistent results. The solder has to be kept molten during the removal of the excess metal, and this involves precise control of the heating of the compressed air supply.

The original air knife systems were complex, difficult to set up and maintain, and also produced very mixed results (Fig. 5.48). However, with con-

Figure 5.47. A close up of the air knife used in soldering to reduce bridging. Courtesy Hollis Engineering.

stant development the newer machines are producing excellent soldering. They are especially useful in providing bridge-free joints on very tightly packaged boards, in particular those using surface mounted components which pass through the solder wave. It has been shown that even multi-leaded components can be soldered bridge free.

The theory covering the use of the air knife points out that a fully wetted joint has the solder strongly bonded to the base metal, while the solder has a very weak bond where it is not wetted. The force of the flowing air will blow off the solder on the nonwetted areas while it will not materially affect the solder coating on the wetted portion of the joints.

Of course as with any of the systems described in this book the ultimate challenge is to select the best system for a particular product, and the recommendation, as always, is to try out any system with the board assemblies for which it is intended to be used.

Test a sufficient number of the most critical assemblies—at least 10,000 joints.

Run the same number of assemblies on the solder machine both with and without the air knife.

Run a similar number of boards on another machine, preferably of a different manufacturer.

Have your QC inspectors report on the results.

Assess the costs relative to the results.

Remember that defective joints, and that includes unwanted short circuits, are the most expensive parts of machine soldering. The additional cost of the machine and the maintenance will be worthwhile if the air knife reduces or eliminates touch-up and possible field failures. Let the results of the tests determine your decision.

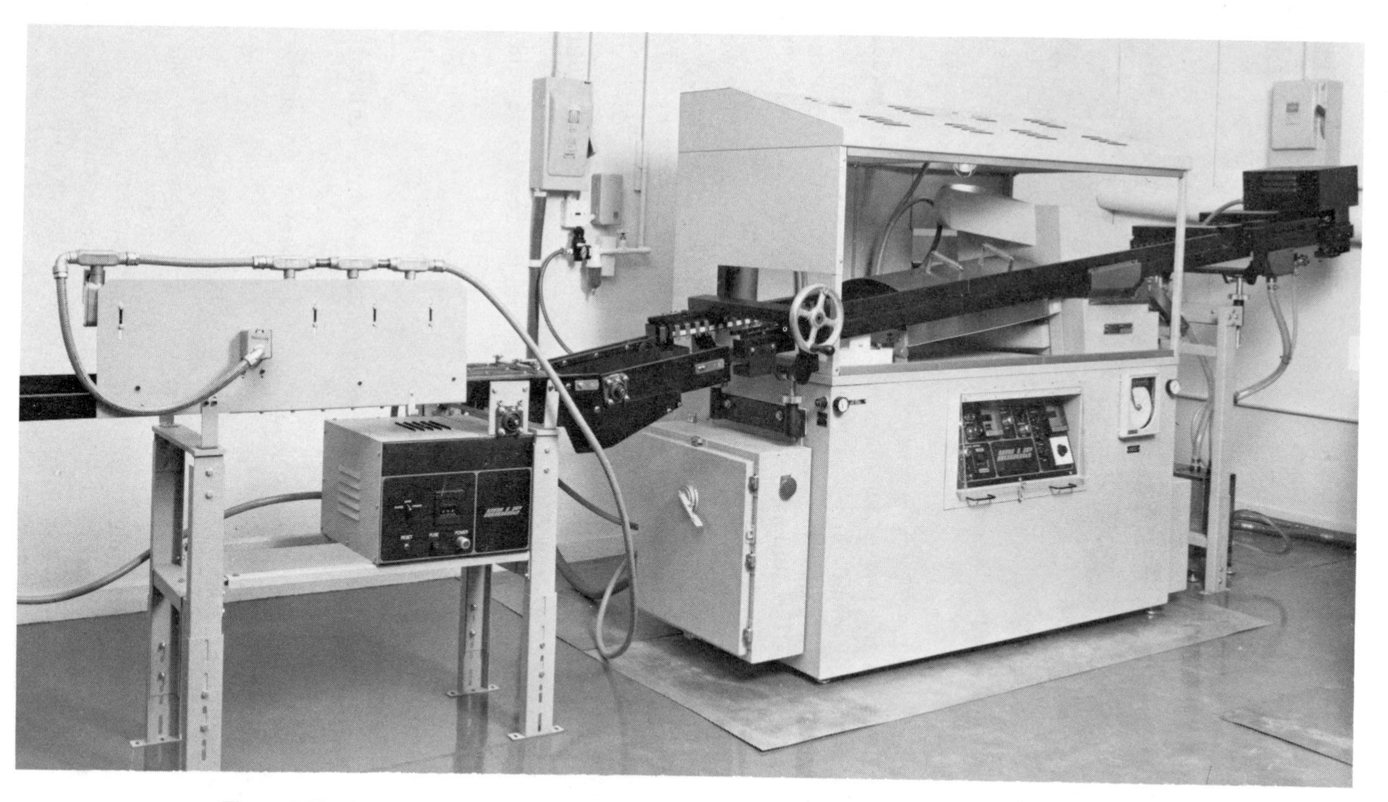

Figure 5.48. An early model of a solder machine using the hot air knife. Courtesy Hollis Engineering.

Static Bath Soldering

This section of the book has concentrated on the solder pot, the solder pump, and the various nozzle styles available on the market. However, there are several soldering systems where little of this applies, although much of the fundamental theory of the effect of the wave shape on the soldering performance will be seen to be relevant. The system referred to is static solder bath soldering. While this method of mass soldering has not gained the popularity of wave soldering, it is nevertheless widely used, especially in Europe and Japan, and has been developed to a high level of sophistication. One popular form is called *drag soldering*. The same methods of fluxing and preheating are used as has already been described. The difference lies in the solder pot and the method of conveying the board over the solder surface. As the name of the system suggests, the solder is not pumped to form a wave, but stays at all times static or stationary in the pot. As the other title of drag soldering infers, the board is pulled or dragged over the solder surface.

The pot is shallow and is wide as the widest board to be soldered. It is also considerably longer than the maximum length of the longest board to be processed (Fig. 5.49).

In operation the board is held in a carrier, which is transported by a chain conveyor over the fluxing and preheating stations. It is then taken at an angle down to the solder surface, where it is floated for a sufficient time to bring the joint to soldering temperature and allow it to become wetted with the solder. The carrier is then moved away from the solder surface at a slight angle, allowing the excess to drain back into the pot under the forces of gravity and surface tension (Fig. 5.50). It can be seen that this part of the process is very similar to that discussed under the heading of Adjustable Asymmetrical Wave, and, of course, is used to produce similar results. The conveying system incorporates controls to allow the various parts of the process to be timed for different types of products. For example, the time

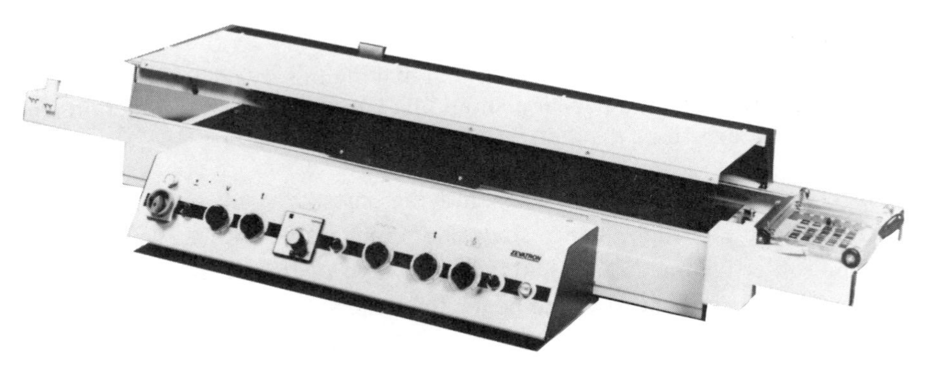

Figure 5.49. A typical small drag soldering machine. The pallet is either passed through the machine or returned to the operator. Courtesy Zevatron GMBH.

Figure 5.50. A pallet in the solder bath of a drag soldering machine. The chains indicate the entry and exit angles of the board in the solder bath. In this machine the dipping length can be automatically varied, permitting boards of differing thermal masses to be soldered without stopping the system. Courtesy Zevatron GMBH.

during which the board is stationary in the solder can be preset, as can the speed of entry and withdrawal.

In other systems the board moves at a constant rate, but the actual length of contact with the solder is adjustable.

Because there is no movement of the solder, some method must be arranged to remove the oxide from the surface prior to the board entering the bath. This is usually done by means of a stainless steel skimmer which is attached to the conveyor immediately ahead of the pallet containing the board to be soldered. Some other methods are occasionally seen such as a spiral surfaced roller which continuously "winds" the oxide off the pot into a dross container.

Drag soldering is capable of excellent results, and should not be dismissed because it is not as commonly seen as wave soldering. As with any system it has certain advantages and disadvantages. These should be evaluated in relation to the particular circumstances for which the soldering machine is required: the type of product to be run, the volume of product, and so on (Fig. 5.51).

The following points may be useful in making this evaluation, but as with any system the ultimate decision should be based on actual production testing and comparison of results:

It is extremely simple, with few moving parts.

It is capable of high quality soldering, comparable with wave machines.

The solder volume is less than that of a wave system of similar capability.

The withdrawal speed of the board from the solder can be adjusted without changing the other processing parameters.

The static pot is claimed to produce less dross than the wave machine.

The drag soldering system can handle boards with leads up to 2 in. (50 mm) long.

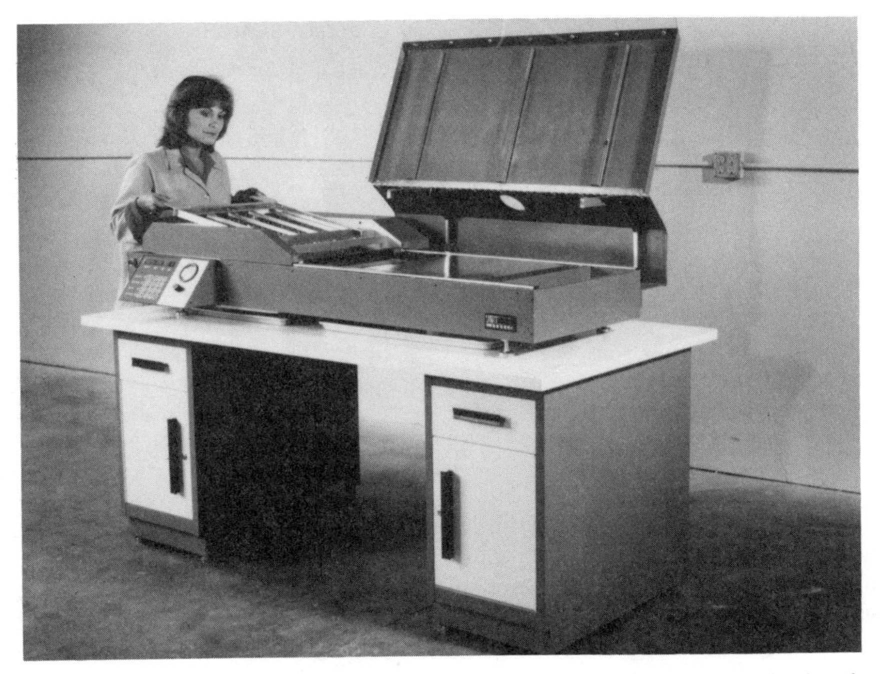

Figure 5.51. A microprocessor controlled drag soldering machine. The solder bath depth makes this also a useful system for timing or soldering long leads. Courtesy Unit Design.

At this time only pallet-type drag soldering machines are available.

The conveyor is generally more complex than in the wave machine.

The maximum length of board that can be soldered is determined by the length of the solder bath. In the wave machine it is virtually limitless.

The simpler drag machines are much slower and have a smaller throughput than a similar size wave machine. Some of the newer, more sophisticated drag machines can compete in output with the wave systems.

Several different versions of the static solder bath system have been developed, especially in Japan, while it is almost universally used as the first soldering station in the solder-cut-solder system described more fully in Chapter 8.

In spite of the term static solder bath, in one system the solder is continuously pumped, but only to remove the dross. Instead of pumping a high wave through a nozzle, the pump is used to gently overflow a large solder lake. This maintains a smooth clean surface into which the board is dipped and avoids the mechanical problems of scraping, or otherwise clearing the dross from the solder surface. The fact that the solder is continuously pumped is said to produce more dross than a totally static pot, although it is doubtful if in practice this is necessarily true, because even in the static pot the solder is subjected to the constant passage of the dross scraper across the surface.

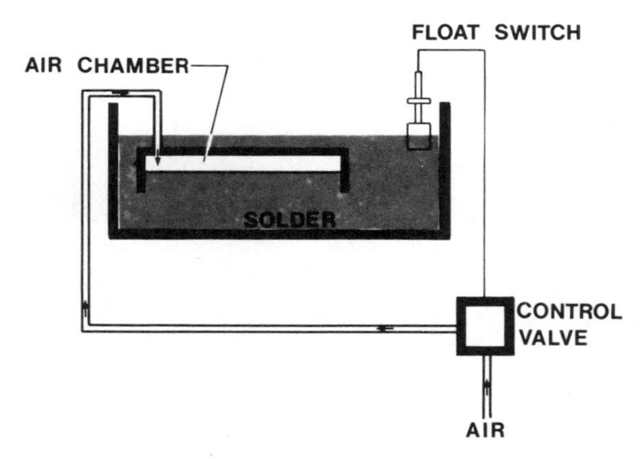

Figure 5.52. A diagram of an air operated solder level control.

With the static solder bath the level of the molten metal becomes much more important than with the solder wave. If the level drops only a few millimeters, this can result in nonsoldered joints, while a similar drop in a wave machine would not affect the height of the wave at all. Of course, the bath's large surface area results in a slow change of solder level.

Almost all these machines are fitted with solder level indicators, many with automatic solder replenishment systems. One vendor uses an ingenious device to maintain a constant solder level. As the solder is consumed, a float switch operates an electrically controlled air valve, which permits compressed air to flow via a suitable pressure reducer into a chamber in the solder bath. The air displaces some of the solder, which consequently raises the level of the bath. This results in a solder level that is claimed to be constant to within 0.3 mm (Fig. 5.52).

As well as several different methods of maintaining a clean solder surface, there are also several different ways of bringing the board to the solder surface. The most common has already been described, where an inclined plane is used to move the pallet into and out of the solder bath.

Another uses hydraulic or pneumatic cylinders to raise and lower the solder pot, while the board to be soldered remains in the same vertical position. The solder pot movement is synchronized with the passage of the boards along the conveyor. In a third system the length of conveyor that passes over the soldering station is carried on cams, which first lower the board down into the solder by dropping one end of the conveyor. They then lower the other end until the board is once again parallel with the solder surface and floating on it. This sequence is then reversed so that the board is removed from the solder at a slight angle, and returned to the original conveyor level. This is a complex mechanical movement, but it is claimed to produce fast soldering speeds with a high quality of soldered joints (Fig. 5.53).

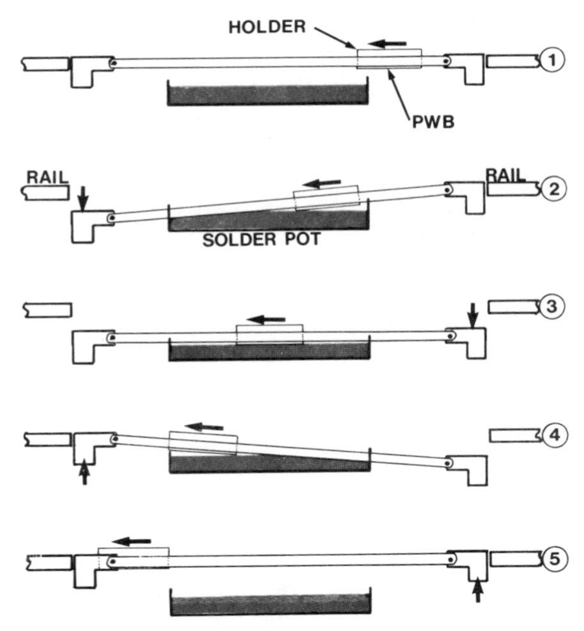

Figure 5.53. A diagram of a complex mechanical system used to pass the board through a drag soldering system.

It can be seen that all these methods of moving the board have one thing in common, and that is to try to produce the optimum speed and angle of withdrawal of the board from the solder, as discussed in great detail in the section on the Adjustable Asymmetrical Nozzle. With the tremendous variety of production requirements there is no single "best" method of performing this function; each system has its own strengths and weaknesses. With all the types of static bath soldering, speed is without doubt the major area in which they find difficulty in competing with the wave soldering machines. As previously discussed some of the newer systems have addressed this successfully, but the very nature of the method of moving the product over the solder surface requires spacing of the boards, whereas the wave system can operate at maximum speed (12 ft or 4 m per minute is not unusual) with the boards placed end to end. This may not be a matter of concern in the choice of a soldering system and is only one of the aspects that should be considered. Excellence in results with perfect joints must always be the prime concern.

Cooling

With all soldering systems the board will take some time to cool to the temperature where the solder has solidified, and until this occurs there is always the possibility that the joint may be disturbed and the quality of the connection jeopardized. In addition, the board remains so hot as to be difficult to handle without gloves for a considerably longer time.

Although this is not often seen as a problem, some manufacturers offer cooling stages fitted immediately after the solder wave. These fall into two categories: the fan type and the air knife. The fan system is by far the simplest and consists of one or more fans mounted under the conveyor with suitable covers to prevent accidental injury or damage from the moving

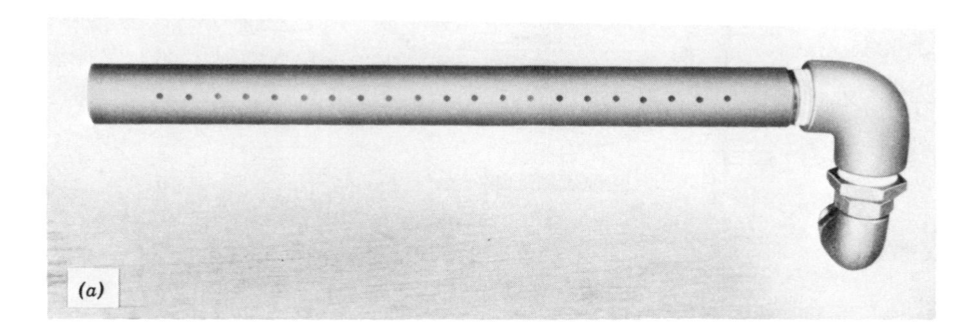

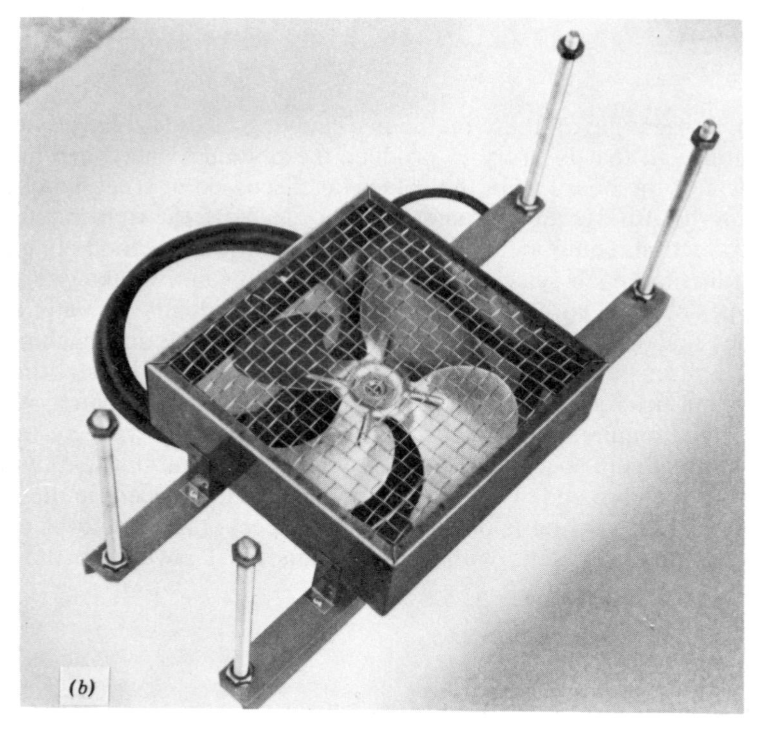

Figure 5.54. (*a*) A cooling fan. (*b*) A cooling air knife. Courtesy Tamura Seisakusho Co. Ltd.

blades. Generally, these are to assist in the solidification of the solder, and therefore the flow of air is gentle so as not to disturb the cooling metal. If it is required to cool the assemblies for handling the air knife is more effective, but it requires a source of compressed air, either from the factory supply or its own blower (Fig. 5.54).

Some specifications forbid cooling the assembly after soldering by anything other than ambient, static air, so before deciding to use a cooling stage all product requirements should be checked.

The reason for this ban is to prevent any possible disturbance of the joint while the solder is still molten. However, a gentle flow of cold air will increase the cooling rate of the solder without any possible jeopardy to the life of the joint. Indeed, there is ample evidence to show that rapid cooling can reduce the thickness of the intermetallic compound and produce a much finer grain structure in the solder. Both of these items will at least theoretically improve the life of the soldered joint.

CONVEYORS

In any mass soldering machine it is necessary to provide some form of mechanical system to move the product through the various stages of the process: fluxing, preheating, and soldering. There are two methods of doing this in use today: the *pallet conveyor* (Fig. 5.55) and the palletless or *finger conveyor* (Fig. 5.56).

The pallet conveyor, as the name suggests, uses a pallet or carrier to actually hold the board, which in turn is propelled by the chains that provide the moving force. The pallet allows the soldering of any assembly within the size limitations imposed by the pallet.

Figure 5.55. A pallet passing down a pallet conveyor. Courtesy Electrovert Ltd.

Figure 5.56. A board supported in a finger conveyor. Courtesy Electrovert Ltd.

The palletless conveyor, on the other hand, is variable in width and moves the assembly by blocks or fingers that are part of the chain conveyor. No other tooling is necessary.

The choice of the conveying system is not straightforward and depends very much on the design of the product, the quantity, and variations in the product mix. The effect of all these parameters on the choice of conveyor will be seen clearly as this section reviews the systems in detail.

The Pallet Conveyor and Pallets

The pallet conveyor can use several different types of pallet, and this flexibility is one of the chief advantages of this system. As well as being used for soldering, the pallet is often also used during the assembly of the board and to move the product through the entire manufacturing area. The pallet may be designed, in these circumstances, to provide facilities other than those specifically necessary for soldering, such as fixturing for cables or connectors, or supporting several parts of the same assembly so that they can all be assembled and soldered at the same time. As long as this form of tooling is designed carefully it is very effective in reducing the cost of handling, and the soldering machine can be integrated into the assembly lines, with the product passing automatically through the soldering station.

This method of using pallets is chiefly confined to very high production rates, where the cost of the pallets can be justified and where a reasonably long production run is assured. (Pallets can cost from $125 upward depending on their complexity.) The pallet may be designed to accept one large board, for example, a complete television receiver, or several smaller boards, which will later be assembled into one complete unit. The boards may be removed after soldering or may stay in the pallet right through to the electrical test.

Where the volume of the product does not justify the manufacture of many special pallet types, but the flexibility of the pallet conveyor is still required, simple tooling plates are often used to obtain most of the advantages of the pallet system, while using a standard pallet frame (Fig. 5.57). Made of steel, fiberglass, tempered hardboard, or aluminum they are routed

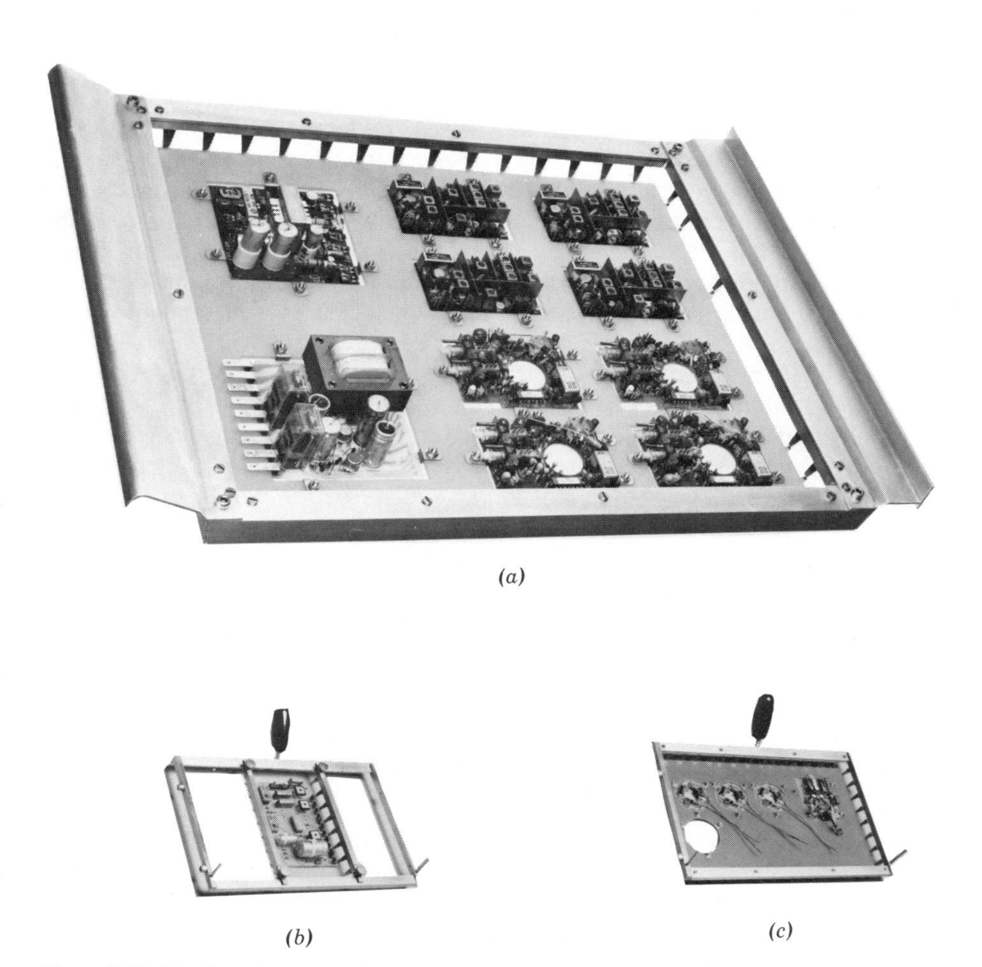

(*a*)

(*b*) (*c*)

Figure 5.57. Tooling plates and pallets. (*a*) Four different type of PWBs in a single tooling plate. (*b*) and (*c*) This adjustable pallet can be used for different tooling plates. Courtesy Zevatron GMBH.

to conform to the outline of the board, which can then be quickly and accurately located in the fixture. As well as supporting the assembly the tooling plate can also be designed to protect those parts of the board which must not come in contact with the solder, for example, gold or other plated contact fingers, or mounting pads that will be used to mount parts later in the assembly process or parts that may contaminate the solder.

These can certainly be protected in other ways, by the use of a liquid solder stop, high-temperature tape, and similar methods, but then the cost of the materials and labor to do this must be added to the soldering costs. Therefore, tooling plates, under the right circumstances, can reduce the overall cost of soldering. In addition, because the board can be retained on all sides, warping is considerably reduced; indeed, large boards can only be soldered satisfactorily in a well-designed pallet.

Assemblies with heavy components should only be soldered in pallets, and even then may require special supports to retain the rigidity of the board during the actual soldering operation. Where several small boards are to be interconnected by means of a cable harness, flat or flexible cable, or flexible circuitry, the tooling plate can be designed to hold all the boards and the interconnecting medium, so that all the joints are made in one pass of the soldering machine. These are only a few of the many ways that the pallet and the tooling plate together provide a very flexible method of handling the assemblies during soldering. Many other possibilities will come to mind when your own requirements are considered carefully.

The pallets must be rigid enough not to bend, warp, or twist under normal working conditions and the heat of soldering. Especially to be avoided is anything that will cause the pallet to tip from corner to corner; this will result in uneven travel and vibration as the assembly moves down the conveyor. This movement may be sufficient to cause disturbed joints, or even worse, joints that have skipped the solder entirely. Although rarely carried out in practice, it is absolutely essential that every pallet is checked by QC at least once a week for mechanical accuracy, as well as to determine if any protective anodizing or other coating requires renewal. Any pallet that is dropped must be taken out of production until it has been inspected and approved as being within tolerance.

Of course any part of the pallet that goes through the system must be made of a suitable material that will not be wetted by the solder or attacked by the flux. In addition, it must not add contamination to the solder pot. Some of the more exotic materials, such as titanium or zirconium are ideal, but the cost is prohibitive except for very small items. Aluminum is often used and is a very good substitute for the more costly materials as long as it is suitably protected by an anodizing or Teflon® finish, so that it does not add aluminum contamination to the solder pot. This finish will have to be renewed from time to time as it becomes scratched or abraded. Steel is an excellent material provided it is protected from the flux and solder; in this case to prevent the solder from wetting the steel. An oxide treatment is satisfactory but, again, will have to be renewed regularly, and steel pallets tend to be very heavy for the operator to handle. From a strength point of view steel is excellent and is the only material for the large tooling necessary for really big assemblies. One pallet designed for a 36 in. (914 mm) machine required a small crane to load it into the conveyor.

For the smaller volume of product there are very many forms of adjustable pallet. These vary from spring loaded fingers on a bent sheet metal frame to grooved plates on a solid casting, which can be adjusted to accommodate different board sizes. They generally retain the assembly satisfactorily, and the chief variation is in the convenience in use, and the ease of loading and unloading. As with all the other parts of the soldering system, try out the pallets with the products that are to be soldered. The standard pallets are made in a wide enough variation to accommodate almost any requirement. For special purposes consider fabricating a simple tooling plate for use with a standard pallet frame (Fig. 5.58). Some other considerations

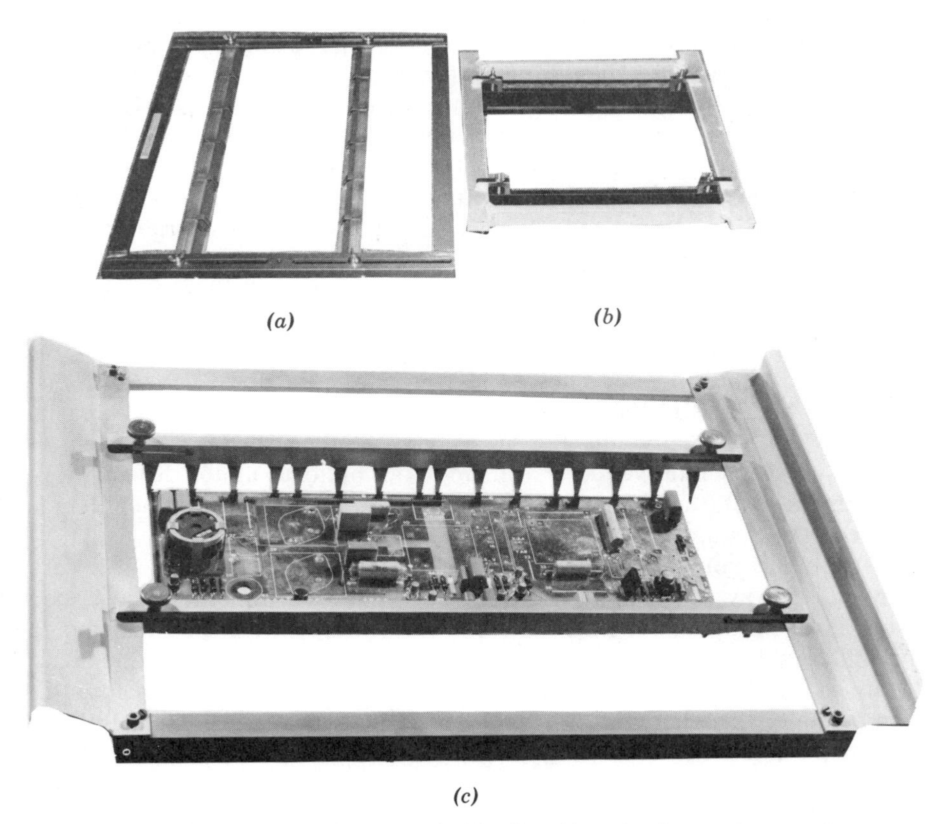

Figure 5.58. Typical pallets. (*a*) Sheet metal with adjustable spring fingers. Courtesy Dynapace Inc. (*b*) Welded aluminum pallet with adjustable side bars. This is typical of the pallets used in a solder–cut–solder machine where the board has to be rigidly contained and accurately positioned. Courtesy Electrovert Ltd. (*c*) Finger-type pallet with a board is position. This pallet is for use with a drag soldering machine. Courtesy Zevatron GMBH.

are the following:

Is the pallet sturdy enough to take the normal shop handling?

Can it be cleaned easily?

Will it be easy to load when it becomes sticky with flux?

Is is easy to adjust for various board sizes?

Is the material from which it is made satisfactorily protected from solder and flux?

Is it easy to handle, not too heavy, with no unnecessary projections?

Does the pallet hold the assembly firmly?

Will the pallet prevent the assembly from warping?

The Palletless or Finger Conveyor

As the name suggests, the palletless conveyor does not require a pallet or indeed any other tooling. The actual conveyor consists of a series of blocks, fingers, or other method of supporting or gripping the board mounted on the chain of the conveyor. They are an integral part of the conveying system; they grip the assembly as it is fed into the conveyor and automatically eject it after passing it through the various stations of the soldering machine. That portion of the conveyor that actually holds the board will, of course, pass through the flux and the solder wave and must neither be corroded by the flux nor wetted by the solder. This means that the correct materials must be used for these components. There is little choice; stainless steel can be used if only mild rosin fluxes are part of the process but will become wetted if more agressive fluxes are involved. Titanium is preferred, and the cost difference is not sufficient to make any substitute worthwhile.

In order to accept assemblies of different sizes, the conveyor rail on one side must be made adjustable, so that the width of the system can be easily altered. This is usually adjusted by a conveniently placed hand wheel, although most machine vendors will supply an optional motor driven width adjustment, with the controls set convenient to the loading station. Details of the palletless conveyor are shown in Figs. 5.59 and 5.60.

Where the volume of boards of the same size is large this is a very convenient way of handling the product, within the limitations described later. If the product mix is such that boards of many different widths have to be soldered, then the conveyor will have to be adjusted each time the board width changes, and the machine has to be emptied of any product before this can be done. Together with the time necessary to make the adjustment, this can result in a considerable reduction in the available machine throughput, and in these conditions the pallet machine, with changeable tooling plates, may provide a more productive alternative.

There are other limitations to the palletless conveyor. The maximum board width that can be run is determined by the rigidity of the assembly. During exposure to the heat of the soldering process there is considerable expansion of the copper clad laminate: about 0.001 in. (0.25 mm) for every

Figure 5.59. A palletless or finger conveyor. Courtesy Electrovert Ltd.

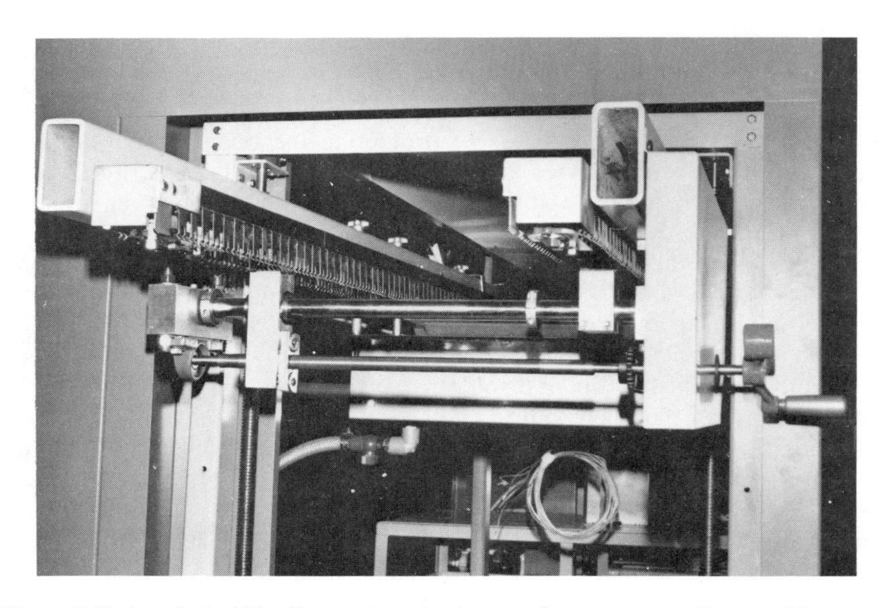

Figure 5.60. A typical width adjustment mechanism on a finger conveyor. Courtesy Electrovert trovert. Ltd.

inch (2.5 cm) of board width. This figure will vary considerably with the board material and direction of the laminate base. At the same time the heat causes a reduction in the strength of the laminate. All this can result in considerable warping or bowing of the assembly. Some of these variations will be taken up by the tension of the conveyor fingers, but only to a small extent. If the board is wide, or the board material very thin, or if the assembly contains heavy components, then some form of stiffening has to be fitted to the board, and this then reduces the economic benefits of the system. Aside from other considerations too much warp in the board can cause it to "dive" into the solder wave, flooding the top surface with solder. This invariably results in scrapping the assembly.

It is not possible to lay down hard and fast rules on the maximum width that can be processed in the palletless conveyor; indeed, some companies deliberately build stiffness into the design of their products so that the convenience of the finger conveyor can be used, even though their boards are of larger than normal size. These stiffeners can take the form of vertical screens between sections of the circuit or large rigid components strategically placed on the circuit board. Generally, 12 in. (30 cm) is taken as the maximum limit, but these figures should only be used as a rough guide. Before committing to any particular system, tests should be run with the actual boards to be processed. Board material, thickness, circuitry, and component layout, and even the use of clinched or straight through leads can all make a difference to the amount of warp that can be expected. It is impossible to forecast what is acceptable and what is not.

With the palletless conveyor, areas on the board that must be protected from the solder, such as gold plated finger contacts, must either be coated with a suitable solder stop, or taped prior to soldering. If the board design permits, slip-on titanium masks can be purchased that simply clip over the gold contacts. All these methods of protecting the board utilize labor in applying the protecting device and removing it after soldering. In this case the pallet system, with the ability to incorporate this protection into the tooling, has some definite advantages.

Where most or all of the boards to be soldered have contact fingers along one edge of the assembly, which must be protected from the solder, the labor involved may warrant the cost of having custom-made masks fitted to the flux and solder pots. These masks, which are usually made of titanium or zirconium, prevent the flux and solder from touching that part of the board that lies between the mask and the adjacent supporting conveyor. This enables boards with, for example, gold contacts along one edge to be soldered without having to protect each board individually. There are some snags; nothing between the mask and the edge of the board will be soldered, so if there are one or two connections in this area they will have to be hand soldered after the assembly leaves the machine. The masks have to be custom-made and require very careful adjustment. They are not cheap since each one has to be individually fitted, and although they can be made adjustable to take care of variations in the size of the area to be masked, the setting-up requires care and considerable skill. However, in spite of all these factors, the use of these masks is well worth consideration under the conditions mentioned above.

Both the pallet and palletless conveyors have their place in the soldering process. The choice is not easy and must be determined depending on the volume, design, and mix of the product. Also, of course, the finger or palletless conveyor will only handle assemblies with parallel sides, and these have to be reasonably accurate. Attempts to run other shapes of boards invariably run into problems and should only be considered as a last resort.

Some users attempt to get the best of both worlds by using the palletless conveyor with pallets as well as with the bare boards. This is usually done by making alternate fingers L-shaped so that the pallet is supported by the horizontal portion of the L. When bare boards are run the alternate V-type finger will grip the edges in the normal way. The L fingers may then shield the board from being soldered close to the edge. This arrangement works extremely well provided that the finger shapes are correctly selected, and the vertical loading, that is, the weight of the assembly and the pallet, is not excessive. The palletless conveyor is generally designed to grip the board with the edges and thus operates with a side thrust. Heavy pallets change this major force to a downward thrust and can cause higher than normal wear of the conveyor. Check with the manufacturer for the maximum weight that the conveyor can handle before instituting this method of operation.

Board Flatness

If the board is not flat when it contacts the solder wave then some areas of the board will be in the solder for a longer time than others. If a board bends down in the center, which is a common problem, then the center of the board may contact the solder wave for two or three times longer than the edges. The contact time with the solder wave is a major parameter in achieving defect-free soldering. It is a product of the width of the contact between wave and board and conveyor speed. If the board warps so that the contact length at the center is twice that at the edges it is equivalent to attempting to solder the edges of the board at double the conveyor speed which is obviously ridiculous.

BOARDS MUST BE HELD FLAT DURING SOLDERING.

If they cannot be kept flat on a finger conveyor then they must be soldered using a pallet or tooling plate. It is very difficult to give exact figures for acceptable flatness as so many factors are involved. A single-sided board will not be very critical in this respect, but a PTH board (especially a multilayer board) may require a flatness within a very small tolerance to solder correctly.

Other Conveying Systems

Conveyors on soldering machines are usually designed to carry the product in one direction only, that is, from the fluxing station to the solder wave. Some systems have been designed to be run with one operator only, and in these the board is loaded at one end of the machine, and after processing is returned to the load station. In these systems the fluxer and solder wave are shut off on one passage of the assembly, and the preheat is usually adjusted by allowing the board to sit over the preheater for a predetermined time. All this permits a very short, compact machine. However, only one board can be in process at any one time, and the output of such a machine is extremely small, compared with a straight through system. For many purposes, however, this is not a problem, and these small machines provide a low-cost, reliable method of mass soldering.

In the low-volume pallet systems, the boards are assembled and then loaded into the pallets for soldering. After this is completed they are removed from the pallet for cleaning, testing, and so on. In the high-volume system where there is a large degree of automation, the bare boards are placed into the pallets and stay there through the assembly, soldering, and sometimes the cleaning and test operations. The pallets are here conveyed by belt or chain conveyor through the entire series of operations, and this is

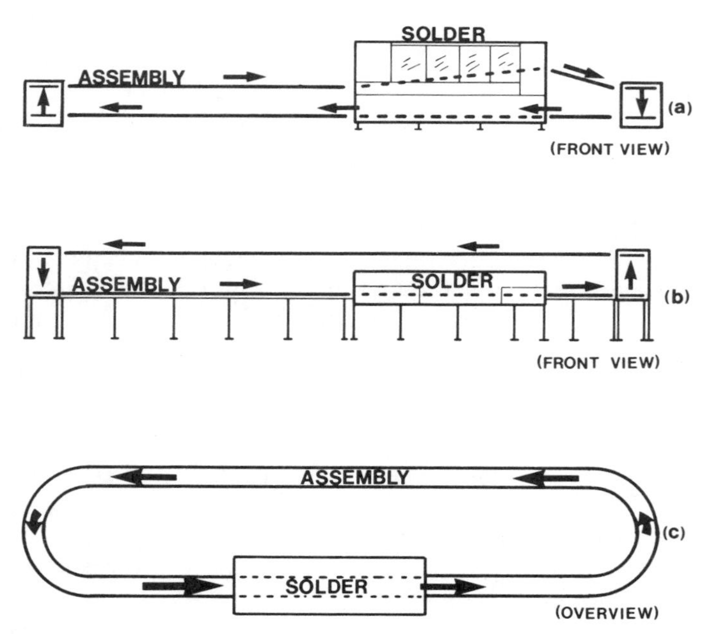

Figure 5.61. Diagrams of some forms of return conveyors. (*a*) Below the solder machine and the assembly benches. (*b*) Above the machine and the assembly area. (*c*) The plan view of a carousel or race course layout.

interfaced directly into the soldering machine. An elevator and return conveyor transfer the empty pallets back to the beginning of the assembly line for reloading with bare boards to start the process over again. The return conveyor may be positioned under the solder machine (Fig. 5.61) or may run over the top of the system. In some designs the conveyor is in the form of an oval or race track, and the pallets are continuously moving through the assembly and soldering area. In this type of conveyor the pallets are sometimes part of the actual conveyor structure and may not be removable.

Where the conveyor also runs through the cleaning operation, it is an advantage to arrange for the pallet to be cleaned at the same time. It is inevitable that flux becomes deposited on it and will have to be removed from time to time. By including the pallet in the cleaning cycle, the task is carried out automatically, and residues will not have an opportunity to build up and cause problems.

In the same way that the pallets become dirty with the flux residues, the fingers of the palletless conveyor will pick up flux, which will become dried and hardened during its passage through the solder wave. If this is allowed to build up, it will eventually prevent the boards from being located accurately and held firmly in the fingers. To prevent this, some form of automatic finger cleaning system should be part of every palletless conveyor (Fig. 5.62). Most finger cleaners consist of a recirculating bath of flux thinners that the fingers pass through during each revolution of the conveyor. Sometimes the solvent is pumped into a wave; sometimes it is applied by means of static or powered brushes (Fig. 5.63). Often a mixture of these methods is utilized.

Figure 5.62. A finger cleaner mounted on a solder machine. Courtesy Electrovert Ltd.

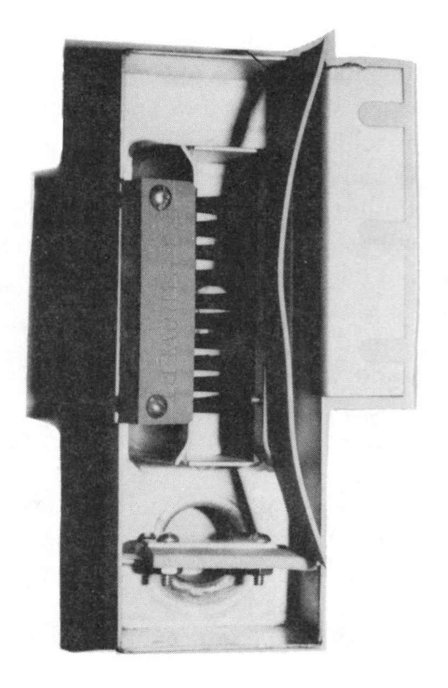

Figure 5.63. A close up of a finger cleaner nozzle. Courtesy Electrovert Ltd.

The finger cleaner is not usually intended to clean off accumulated, hardened flux, and the fingers must be clean to start with. The finger cleaner must be in continuous operation while the conveyor is running. If the fingers become caked with dried flux due to failure to operate the cleaner, they must be scraped or brushed to remove the excess material before putting the cleaner back into operation. The solvent reservoir is usually mounted on the machine and will contain a filter to remove any solids that may be picked up during soldering and that could damage the pump. This filter must be cleaned regularly and the solvent changed when it becomes too dirty to do the job (Fig. 5.64).

The materials of the pump and the piping are chosen to be compatible with the solvents normally used with fluxes; no other solvents should be used. Occasionally pumps fail, and piping is reported to rot out in the field. This is invariably found to be due to the use of one of the stronger industrial solvents. Although they may not cause immediate damage, they can eventually cause the pump rotor to swell and jam or the pipes to disintegrate. The standard thinners as recommended by the flux vendor will do an excellent job of cleaning and will not harm any of the equipment.

Conveyors have to operate smoothly over a wide range of speeds and be set accurately at any predetermined speed; the conveyor speed is one of the major soldering parameters. Drive is usually from either a variable speed DC motor or from a fixed speed AC motor with a variable speed mechanical drive of some sort. Both these methods provide a perfectly satisfactory drive system, and the choice is really one of convenience. The most important

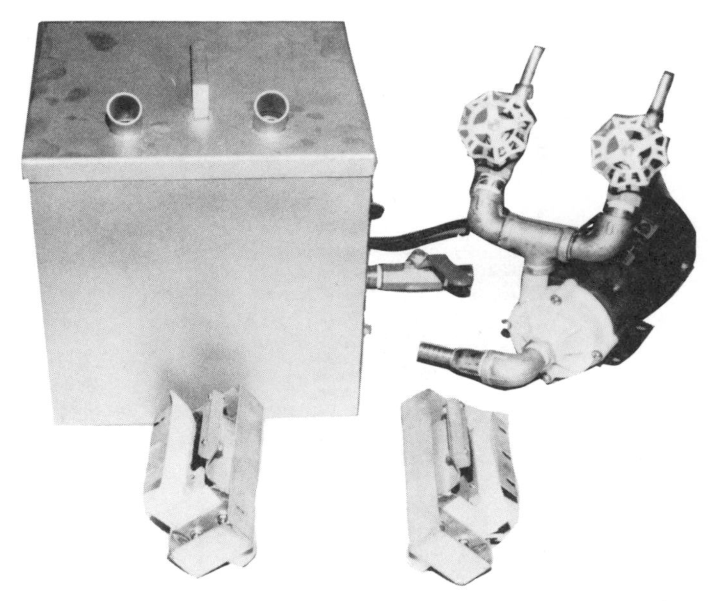

Figure 5.64. The parts that together form a finger cleaner system: the pump, the nozzles, and the supply tank. Courtesy Electrovert Ltd.

factors are that the drive should operate smoothly, without changes to the conveyor speed, and that when it is required to change the speed this can be done easily and to accurately repeatable settings.

The DC drive can have the speed control mounted remote from the mechanical part of the conveyor drive, and especially with the feedback type of controller will provide excellent speed stability over a wide range of settings. The variable speed mechanical drive, on the other hand, has to have the control adjacent to the drive mechanism, which may not necessarily be convenient for setting up the machine (Fig. 5.65). The AC drive also tends to be somewhat speed variable, depending on the loading of the conveyor. If the frictional losses on the conveyor becomes excessively high, usually because of inefficient or irregular maintenance, the speed may be slowed down without any indication that this has happened. This is not to infer that the AC drive is not perfectly satisfactory under all normal conditions, but does emphasize the need for regular and careful maintenance. All conveyors should be fitted with an overload clutch to prevent damage to the machine, the product, and the operator in the case of any jamming of the system. It is all too easy for fingers or clothing to become entangled in spite of the safety guards fitted. The clutch can be fitted with a switch that will cut off the power to the conveyor automatically if the clutch slips more than a few degrees. The power will then stay off until the fault is resolved and the circuit reset.

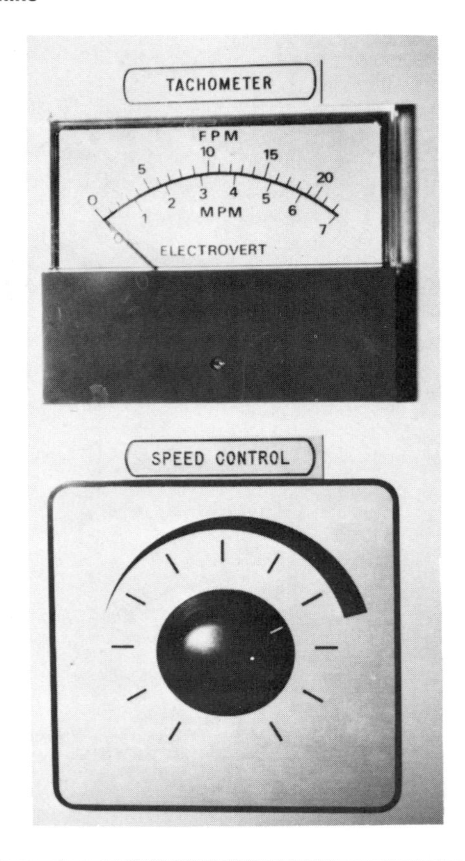

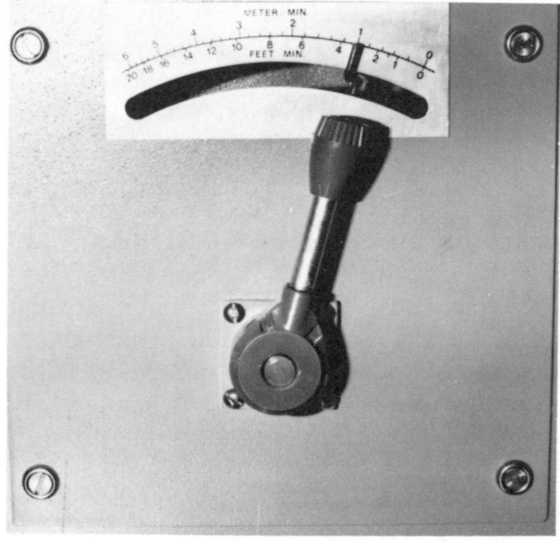

Figure 5.65. Various analogue conveyor speed controls and indicators.

All speed controls must have some form of calibration so that the speed settings can be repeated accurately according to the predetermined requirements of the products to be soldered (Fig. 5.66). The calibration can take many forms, from a scale against which the variable speed drive control handle moves, to the sophisticated digital system of the DC feedback drive. Whichever type is used it must be checked regularly. This is very simply done by timing the passage of a board or pallet over a known length of the conveyor. It is convenient to place permanent marks on some portion of the structure for this purpose. The results should be recorded as a matter of routine as one of the possible variations in the process.

In the section on nozzles the effect of the modern wave shapes on the soldering results was discussed in detail, including the essential part played by the use of the angled conveyor. Most soldering machines today are fitted with these conveyors; indeed, there are few made with a totally horizontal system, both in the pallet and palletless forms.

Where only one nozzle type is used, with boards of a similar design, a

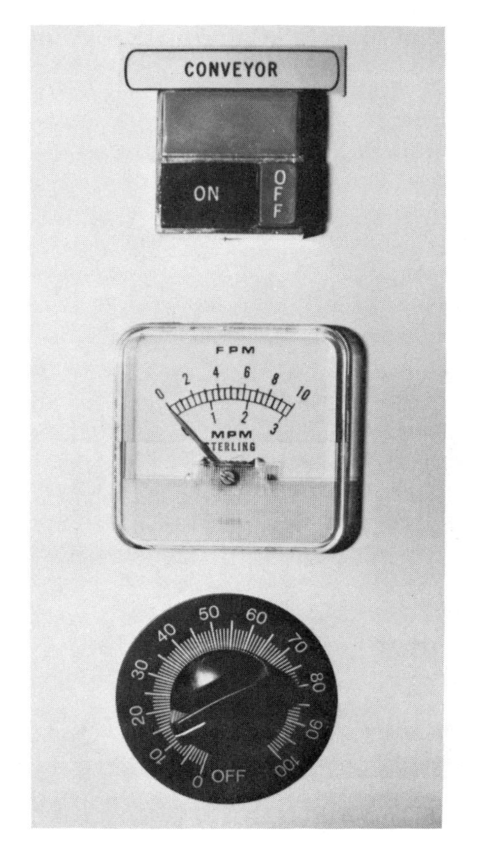

Figure 5.66. A simple analogue conveyor control system usually found on the small soldering systems.

fixed angle conveyor as recommended by the machine vendor is completely adequate. The angle is usually chosen to be somewhere between 4° and 6°. Where it seems likely that the nozzle will be changed for one of another type at some time, or that assemblies of radically differing designs will be soldered, then a conveyor should be fitted to the machine that is capable of being varied in its angular relationship to the solder wave. This can be achieved in different ways. The simplest method involves the use of a spanner to make the adjustment, and a maintenance technician to carry out the task and to assure that the conveyor is correctly aligned once the new angle is established. This is not an unreasonable task if the adjustment is only made occasionally. If it has to be carried out frequently, then the machine down time cannot be tolerated, and a system that can be set by the operator is preferable. In one method the angle can be changed simply by turning a handwheel until the required angle is set as indicated on a scale on the machine (Fig. 5.67). Thus, the angle cannot only be changed quickly, but it can be set accurately without any external measurements being required. Note that the solder pot height will normally need to be changed each time that the conveyor angle is readjusted. Both the conveyor angle and the solder pot height adjustments can be obtained in power operated versions which offer the ultimate in convenience and speed in making these machine settings. To summarize then:

Use a pallet conveyor when the volume of any particular board is small and many different types and sizes have to be soldered.

Use a pallet when boards
- Are irregular in shape.
- Do not have parallel sides.

Figure 5.67. The mechanism used to vary the conveyor angle. Courtesy Electrovert Ltd.

Need masking for gold fingers.
Are large and have heavy components.

When using a palletless conveyor
Board warping must be minimal.
Board tolerances must provide reasonably parallel sides.
Components near the edge may not solder.
Components must not extend beyond the edge of the board where held by the fingers.
The assembly must not exceed the weight specified by the machine manufacturer.
The conveyor must not be overtightened on the board.
Allow for the thermal expansion of the board.

With all conveyor systems, perform regular maintenance, adequate lubrication, and good housekeeping.

CONTROLS

Each section of the soldering machine requires controls to be able to set up the operating parameters. In addition, there may be controls to take care of some of the monitoring requirements, such as solder height or flux density, and there may be indicators to show the status of the various parts of the machine. There are so many possible variations that it is quite impossible to describe all the systems that may be found and the different combinations available. These range from the simplest, which can be a few switches and a temperature controller, to the most sophisticated, which can offer an almost totally automatic process with computer control and no operator intervention. All the more commonly used control systems will be described in detail, together with some probable combinations.

Temperature Controls

Temperature controllers are the most common instruments to be found in soldering machines (Figs. 5.68 and 5.69). All systems must have control of the temperature of the molten solder, as well as the preheaters, and occasionally the flux pot.

The temperature of the solder pot must be controlled to at least plus or minus 5°F (3°C) and preferably to tighter limits. Although the wider limits have been used satisfactorily for many years there is a growing trickle of evidence that many factors considered unimportant in the past may well contribute to the small percentage of faulty soldered joints that must be eliminated if the objective of zero defect soldering is to be achieved, and this must be the target for the industry. Temperature controls—no matter which

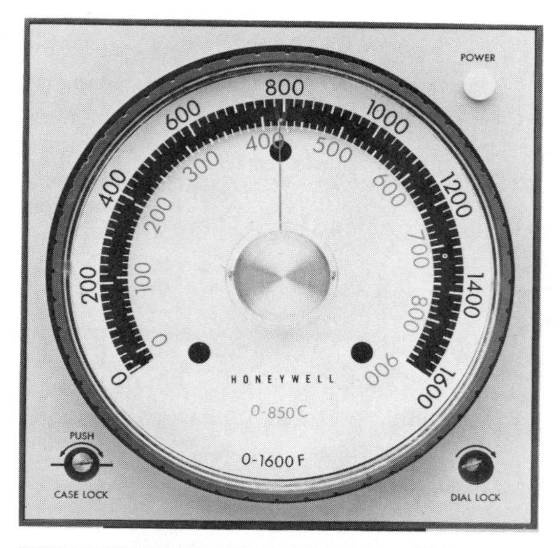

Figure 5.68. An analogue temperature controller and one with digital setting switches.

type are used—must have the following features:

An adjustable set point so that the pot temperature can be changed readily to suit the soldering parameters

A continuous monitor of the pot temperature, with a readout showing the temperature and the set point so that the variation can be easily determined; a better alternative is a single read out of this difference

A fail-safe mode so that the pot temperature will not run up in the event of a thermocouple failure

Figure 5.69. A totally digital temperature controller. This avoids errors in setting and reading the temperature. Courtesy Gulton Industries Inc.

It would be a mistake to assume that just because a controller is fitted to the solder pot the temperature will always remain exactly the same, and the various variations, although small, should be known and the reasons for them understood. If a thermocouple and a temperature recorder are used to monitor the pot temperature for an extended period of time, curves such as those in Fig. 5.70 will be obtained. On switch on, the temperature rises as the full power of the heaters is used to melt the solder. The measured temperature almost levels out for a short-time as the heat is used to change the state of the solder from solid to liquid, and then rises once more to the set point of the controller. There is a certain amount of overshoot depending on the type of controller, and the care with which it has been set up by the manufacturer during calibration. With the examples shown this is a small amount. The pot then stays at the set temperature with only a small cycling variation as the controller activates the heaters. With the on–off controller this will be about plus or minus 5°F (3°C). With the time proportional controller the cycling variation is less than half that amount. It is not practical to look for tighter control, and because of other variations in the solder pot it is doubtful if it would offer any practical advantage.

When the solder pump is switched on there is an immediate variation in the solder temperature. This is not shown to any extent in the curves of Fig. 5.70 because these were taken in a large pot with a big solder content. The temperature will fall as the solder heats up the cold nozzle, and with a small pot and a large mass of metal in the nozzle quite large variations have been observed. There may be some temperature "layering" in the pot, although there is little evidence to show that this occurs under normal circumstances. It is probable that there are several factors that together cause this initial

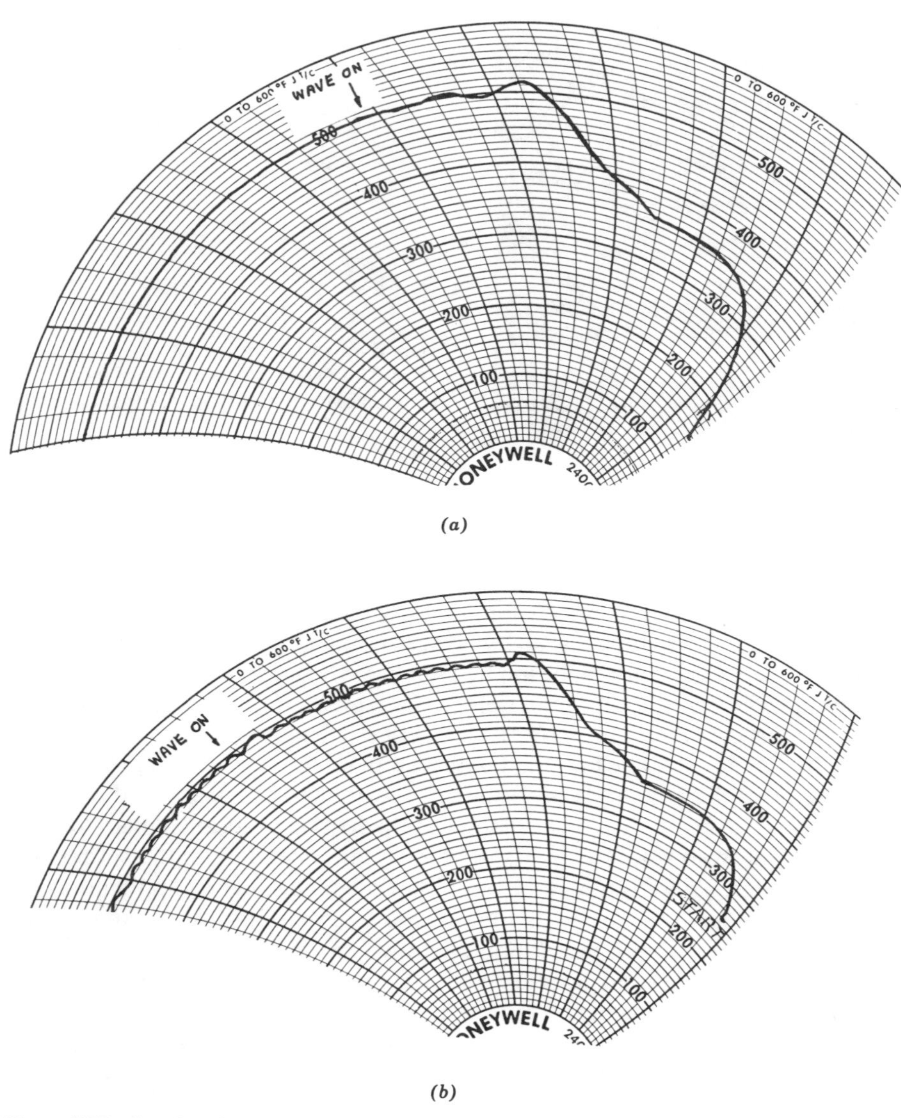

Figure 5.70. Graphs of solder pot temperature using (*a*) a time proportional temperature controller and (*b*) using the on–off temperature controller.

drop in temperature. It will then take from several seconds to several minutes for the pump to circulate the solder sufficiently to bring everything up to a steady temperature once more. The time it takes will depend on the size of the heater, size of the pot, position of the thermocouple and so on. This temperature variation is seldom a problem, but its magnitude should be known and checked from time to time. Any increase may indicate that one of the pot heaters has failed. This also can be of concern if the pump is switched off for any length of time during processing, either automatically by the passage of the board, or by the operator for any reason. Know what the variation is and run a few tests. If an unacceptable change occurs, then the operating instructions for the machine must give the minimum time that the pump has to be run before any product can be put into the system.

In order to accelerate the heat up from cold, some manufacturers fit additional heaters that are only used until the temperature comes within the set point of the controller, when smaller powered heaters are used to maintain the operating temperature. The single set of heaters is usually totally adequate, when coupled with a time clock to switch the heaters on at a predetermined time before the machine is required for production. Occasionally, if a high volume of product is processed, using pallets with a high thermal capacity, or containing components that will absorb a great deal of heat, the pot may require additional heaters in order to maintain the correct solder temperature. Although this is not a common problem it does occur occasionally and should be considered if any of the above conditions apply.

As mentioned above, it is customary to use a time clock to switch on the pot heaters, and also sometimes the preheaters, in advance of starting production. These can be obtained in many styles. Some will omit Saturdays and Sundays; some can be found with backup spring power so that supply failures will not upset the schedule. This latter feature is well worth the small additional cost (Fig. 5.71). Some users operate their solder pots all night at a low temperature on the assumption that this uses less power than shutting down and heating the pot again in the morning. This is not so, and for the sake of saving power the supply should always be switched off when not in use, always provided that there is sufficient time to get the pot back to temperature before it is required again. Most machine manufacturers supply some type of cover for the solder pot when it is not in use. If not, one can easily be made from one of the metallized heatproof fabrics sold for insulating furnaces and other heating equipment. A cover should always be used when the pot is not in actual use; it saves power and prevents the possibility of solder splashes from the pot when it is heating up. This phenomenon is fortunately rare, but trapped air expanding in the pot can force molten solder at a high velocity through small cracks in the still solid solder crust. Of course, there is no possibility of this occurring once all the solder is molten, and a cover prevents this unlikely occurrence from causing damage to the machine or harming the operator.

There are some other functions that the temperature controller can and sometimes does perform, depending on the type and model used. It can provide additional contacts that will operate when the measured temperature is within a predetermined percentage of the set point. This can be used to prevent operation of the solder pump before the solder is molten. It can give an audible warning or other signal if the temperature exceeds the set point. It can warn of thermocouple failure. The more sophisticated types can

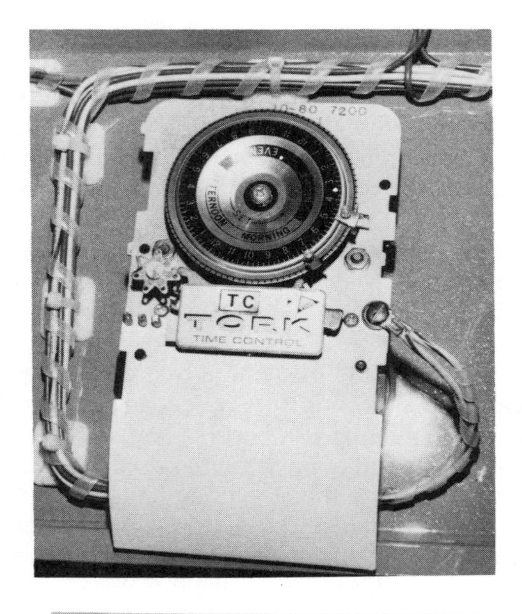

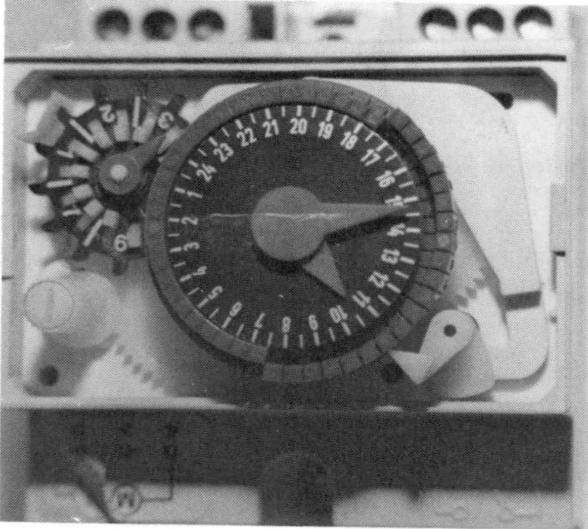

Figure 5.71. Two typical time clocks used to control the switching of solder pots.

be interfaced with a microprocessor, will give remote readings of temperature, and can have the set point changed from another control position or through a master computer. When purchasing a machine it is well worth discussing the various options available, so that the maximum accuracy and convenience are obtained for the particular set of conditions under which the machine will operate.

As discussed in the section on Preheat, this station is an important part of the soldering process, and requires a similar control of temperature as in the soldering stage. Here again as the industry looks for a lower and lower failure rate, it is being found that the control of the preheating appears to be more critical than had been previously considered. This is not an easy area to control because there are so many variables that affect the temperature of the assembly: the board material, the type and quantity of the components, and so on. For a long time, it was believed that these other factors made very accurate control of the preheat temperature unnecessary, and if the less critical products are to be soldered this is probably true. Therefore, it is common to find that the preheaters are often controlled by the simple appliance-type controls as are found on the kitchen stove (Fig. 5.72). These appliance controls are not calibrated in temperature, but rather with some form of arbitrary numbers. It is possible to use these calibrations to reset the controller to approximately the same predetermined setting, or to increase or decrease the temperature by some rather ill-defined amount. If the highest-quality soldering is not required, or if the product is not tightly packaged, or otherwise difficult to solder, this is probably adequate. If, however, the aim is to accept nothing but the best in soldering, this is one area where consistence in soldering machine parameters demands more accurate control with improved repeatability.

Most machine vendors will provide more accurate controllers that operate in exactly the same way and to the same accuracy as those used in the solder pot. These will assure that the preheaters can be set to.the correct temperature with a high degree of repeatability. They are well worth the additional cost if zero defect soldering is the objective.

Flux Density Controllers

As discussed in the section on Flux Density Control, if the fluxing system used is anything other than a total loss type, then the density of the flux has to be monitored and the necessary additions of thinners made on a regular basis. This can be done effectively using an ordinary hydrometer, but it becomes time consuming, and relies heavily on the memory and integrity of

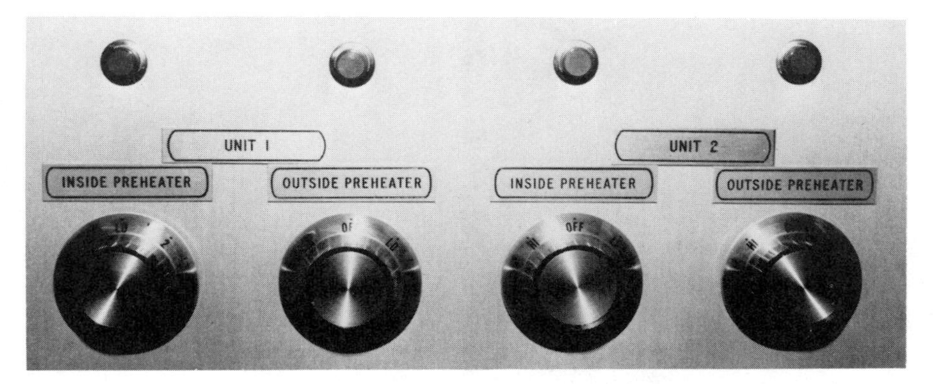

Figure 5.72. The kitchen stove–type of power controller frequently seen on preheaters.

the operator. The automatic system as described in the above section is a popular and effective method of ensuring that the necessary control is done correctly. As the controller is very much an integral part of the fluxer, it was described fully as part of the fluxing process and is only referenced here for convenience and completeness of the control systems.

Conveyor Speed Controls

The mechanical functioning of these was discussed in the section on Conveyors and reference was made to the various ways of controlling and displaying speeds: the mechanical variable speed drive with a scale under the handle, the DC drive, and the more precise and accurate feedback tachometer-type drive. Of these, only the tachometer reads out the actual speed of the conveyor. The remainder read either the position of a control knob or the voltage applied to the drive motor. They are reasonably satisfactory if the loading of the conveyor remains constant and the maintainence on the conveyor is regularly carried out. The obvious danger is that for some reason the loading may change, and if the change is not sufficient to be obvious, then boards may be soldered at some speed other than that which was determined to be the optimum for that product.

This also points out the need to check the calibration of these controls, no matter of which type, on a regular basis. Figures 5.65 and 5.66 show a selection of conveyor speed controls and readouts, and Fig. 5.73 shows a digital speed controller, which offers one of the more accurate speed setting and readout devices. The use of the digital input switches allows no errors in setting the required conveyor speed, and the readout shows the actual speed to two places of decimals. This type of controller can also be obtained with suitable inputs and outputs to interface with a controlling computer.

Figure 5.73. A digital conveyor speed controller. When used in a true feedback system this provides the most reliable control of conveyor speed.

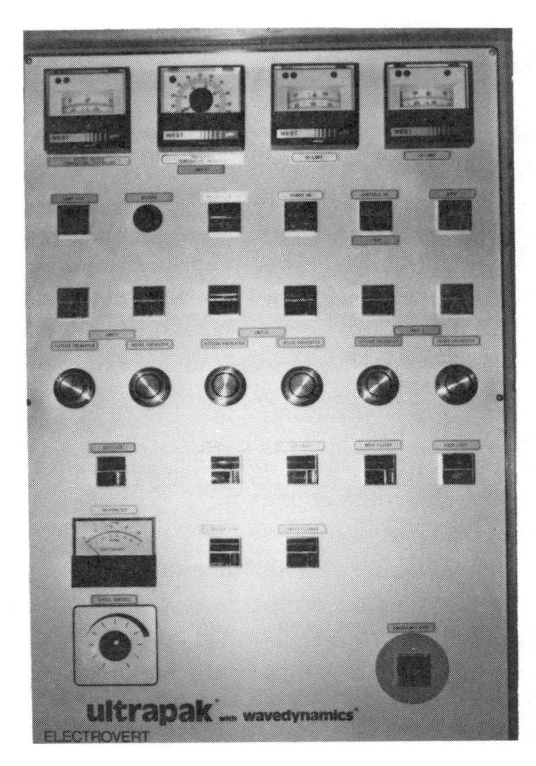

Figure 5.74. A complex control panel. Courtesy Electrovert Ltd.

Whichever type of speed controller is fitted to the solder machine the objective is to be able to set the speed accurately to any predetermined setting and to ensure a high degree of repeatability. Once set, the speed should remain constant with all possible variations in product load and other operating parameters.

Auxiliary Controls and Indicators

There are many levels of sophistication and complexity in the controls and indicators that can be fitted to the soldering system (Fig. 5.74). Those described above are the minimum necessary to ensure correct operation of the process; others are fitted for greater convenience, better control, reduction of labor, or as a warning of failure of some part of the machine. Most machine vendors will supply any of these as options when the machine is ordered. This applies especially to the larger systems which are in any case custom made from standard modules. Consider carefully which will be cost effective for your own particular set of conditions. The following is a reasonably complete list of those commonly supplied:

Warning lights to show any failure of the heaters in the solder pot or in the preheating section; an ammeter is sometimes fitted instead but is not so convenient to use

Solder level warning lights, which are usually part of an automatic solder replenishing system

Temperature recorders on the solder and preheater stations, either the circular or chart type

Counters to show the number of boards processed

Timers to indicate the number of hours the various parts of the system operate

Sensors on the conveyor to monitor the position of the pallets or boards; this information is usually used to switch the fluxer and wave off and on

Readout of conveyor angle

Readout of conveyor width, usually used with power operation and remote control of the palletless conveyor

Failure light to indicate clutch slip

High and low level indicators for the flux container

Over temperature and under temperature warning for the solder pot, either visual or audible

Readout of board surface temperature after preheat

Solder wave height

Computer Control of the Solder Machine

With the increasing use of the computer to control machines, it is only natural to consider the possibilities open to the use of the computer in controlling the soldering process. It is sometimes assumed that adding the computer will immediately eliminate all the soldering problems; unfortunately, that is not likely to occur.

As will be discussed in detail in later sections of this book, the soldering process cannot be considered to be only the actual operation of soldering. The design of the assembly, the solderability of the components and the PWB, the assembly operation, and the handling of the parts are all part of the soldering process, and the computer controlling the soldering machine has no ability to modify any of these factors. What therefore can be expected of a machine with computer control? The answer of course is the elimination of many of the variables in the operation of the machine, especially those that can be attributed to the operator or the setup.

Computer control can be achieved by two methods. In the first, a standard manual machine can have all of the normal control systems interconnected with a central controlling computer. For example, the output of the temperature controller on the solder pot will be transmitted via a suitable converter to the input of the computer, while from the output of the computer will come a signal that can change the set point of the controller. In a similar manner, all the other controls and switches will indicate their status to the computer and can have that status changed by signals from the computer. This, of course, will allow the machine to be set up from a remote terminal, from a card reader on the machine, or by reading a bar code on the PWB. The possibilities are almost limitless. In a similar manner, the performance of the machine will be able to be monitored from a remote terminal, possibly on the desk of the ME or supervisor responsible for the process, with warning of any excursions beyond the specified limits. Note, however, that it is the machine that is being controlled, not the process.

The computer will not know if the incidence of frosty joints is increasing or if dewetting is becoming a problem. There is no doubt that sensors will be developed that will be able to do this, but all too often the problem is caused by factors outside the actual machine operation, and if the computer changed machine parameters in order to achieve a cure, the result would be chaos.

However, this does not mean that computer control is not an effective tool. Even with the system described many process variations will be eliminated. The second method of computer control, which is nothing more than a logical step forward, has even more to offer (Fig. 5.75).

The computer can do much more than just replace the operator. It can do things that it is just not possible for the human operator to achieve. It has the ability to calculate at high speed, while simultaneously controlling, storing,

Figure 5.75. A modern soldering machine in which all the functions are under the control of a central computer. Courtesy Electrovert Ltd.

Figure 5.76. A close-up view of the controller in a computer controlled soldering machine. Courtesy Electrovert Ltd.

and recording information. It can only do these things if the correct information is available in the right format and if the mechanical design of the soldering machine is adequate to allow the computer to carry out the controlling functions. For example, the computer can control the height of the solder wave, but only if it can measure the height required and the actual height obtained, and there is a suitable drive mechanism on the solder pot jacking stand and the solder pump. The second method of computer control of the soldering machine therefore begins with a machine specifically designed to make full use of computer control.

A typical machine with full computer control has few external knobs, dials, or switches. It is totally automatic, does not require an operator, and therefore these items are superfluous. Controls are only required for the initial programming of the system, maintenance, and calibration (Fig. 5.76) and (Fig. 5.77). In this system, control will be completely in the hands of the computer. For example, there will be no temperature controller. The computer will sense the temperature of the solder pot directly from the thermocouple, check in its memory for the correct set point for the particular product being run, and control the pot heaters to arrive at that setting. If the bar code reader at the input to the machine reads from the next assembly that it requires a different temperature, then the new set point will be pulled from memory, and the entry of the board prevented until the pot reaches the new temperature. Rather than monitoring the temperature of the preheaters,

Figure 5.77. The microprocessor control panel on a simple drag soldering machine. Courtesy Unit Design.

which has been shown to be only an indirect measurement, the computer will use the actual temperature of the board surface and control the preheaters to produce the correct temperature regardless of the variations in the various board parameters, such as component loading, board emissivity, and so on.

Built-in sensors will allow the computer to know exactly the position of every board in the machine and switch on and off the solder wave and fluxer accordingly. It will be possible to have more than one fluxer with the correct one being brought into operation automatically from data read by the computer from the part to be processed. The solder wave will be controlled in height according to the measured position of the board. The addition of the computer to a machine specifically designed for that purpose opens up many possibilities for improving the capability and reliability of the soldering process.

With all of the control centered in the computer, almost any requirement can be satisfied through software changes. The following is a short list of some of the more obvious practical possibilities:

Automatic control of conveyor speed from data stored in memory
Automatic width adjustment of the conveyor from built-in sensors

Conveyor angle adjustment from data in memory, with automatic compensation of solder and flux pot heights

Automatic control of flux foam or wave height from board position sensors

Automatic flux density control, and flux dumping and refill

Preheaters controlled from a measurement of the board surface temperature, with automatic compensation for conveyor speed changes or thermal characteristics of the board

Automatic control of wave height and solder level

Continuous recording data, for example, number of boards run, type, serial number, hours of machine use, warnings of maintenance required, flux and solder usage, power consumed, and down time

The record keeping capability alone will prove of immense value in achieving the goal of zero defect soldering, since any changes in the quality of the soldering will be able to be checked instantly against the data for any variations in the process (Fig. 5.78).

With development there are more exotic controls that may become practical: automatic and continuous monitoring of the flux for contamination and the solder for composition, or perhaps checking the solderability of the parts before or as they are soldered, with some form of process compensation

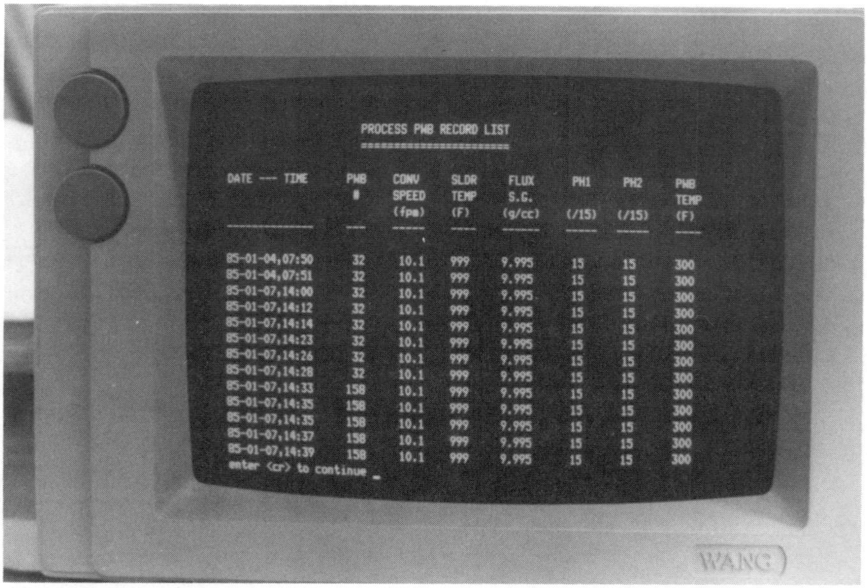

PROCESS PWB RECORD LIST

DATE — TIME	PWB #	CONV SPEED (fpm)	SLDR TEMP (F)	FLUX S.G. (g/cc)	PH1 (/15)	PH2 (/15)	PWB TEMP (F)
85-01-04,07:50	32	10.1	999	9.995	15	15	300
85-01-04,07:51	32	10.1	999	9.995	15	15	300
85-01-07,14:00	32	10.1	999	9.995	15	15	300
85-01-07,14:12	32	10.1	999	9.995	15	15	300
85-01-07,14:14	32	10.1	999	9.995	15	15	300
85-01-07,14:23	32	10.1	999	9.995	15	15	300
85-01-07,14:26	32	10.1	999	9.995	15	15	300
85-01-07,14:28	32	10.1	999	9.995	15	15	300
85-01-07,14:33	158	10.1	999	9.995	15	15	300
85-01-07,14:35	158	10.1	999	9.995	15	15	300
85-01-07,14:35	158	10.1	999	9.995	15	15	300
85-01-07,14:37	158	10.1	999	9.995	15	15	300
85-01-07,14:39	158	10.1	999	9.995	15	15	300

enter <cr> to continue _

WANG

Figure 5.78. This picture shows the data automatically produced by a computer controlled soldering system. Courtesy Electrovert Ltd.

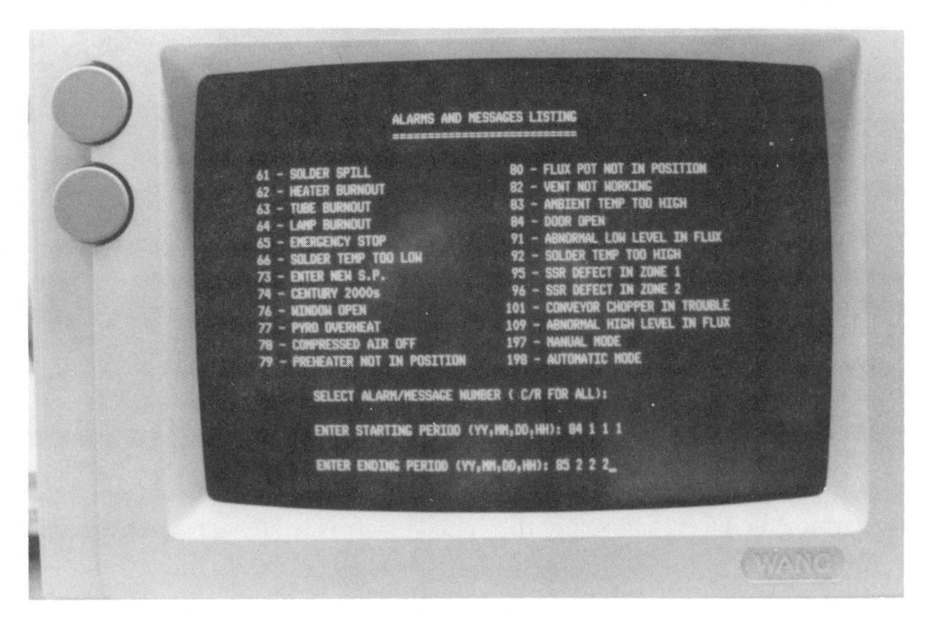

Figure 5.79. The monitor in this picture displays the many points in the solder machine which are monitored in a fully computer controlled system. Courtesy Electrovert Ltd.

when solderability is reduced. This, of course, is all in the future, but soldering machines modified to the first method described above are already in use, and new models incorporating all or some of the capabilities listed will soon be reaching the market.

Machines of both types will reduce the variability of the process and, if used correctly with solderable parts and properly designed assemblies, will eliminate the high cost of faulty solder joints. Maintenance of these machines will become a much more important factor than with the less sophisticated systems and will have to be recognized as an integral part of the soldering process (Fig. 5.79).

Putting It All Together

The various parts of the soldering machine have been described in detail, with the advantages and disadvantages of the different types of systems. Now the total machine will be considered with the requirements for setting up and operating.

It is not often possible to arrive at the optimum detailed design for each and every part of the machine. Because of cost, availability, space, or many other reasons some compromise is usually necessary, and the overall capability of the total machine must be the main consideration. For example, it may be worth accepting the inconvenience of manual operation of the solder pot height to be able to buy the improvement in process reliability offered by automatic flux density control.

These factors have to be considered relative to the special conditions that exist on the shop floor, the product to be processed, and the many other items that have been reviewed in the preceding pages. It is not possible in any book to define all the factors considered in making these decisions, but there are some general comments that may be useful:

It is generally more cost effective to buy process reliability than convenience.

Reliable, well-trained operators can produce excellent results with simple machines.

It seems inevitable that the size and volume of product to be processed will always increase; do not buy too small a machine.

Remember to include the cost of the ancillary equipment: shelves, tables, washing facilities, and so on; they must be provided for the process to run smoothly.

Ruggedness and reliability are as important as sophistication in design.

The smaller soldering machines are usually bench mounted or stand on separate bases that are supplied by the machine vendor, and usually provide suitable storage for tools and supplies. Whichever is used the support must be stable and rigid and free from any wobble or other movement. The weight of the solder alone is usually considerable. Not only is absolute steadiness required to maintain soldering quality, but it will prevent any possibility of burns from molten metal that might splash or spill due to accidental movement of the machine.

The machine should also be mounted in such a position that there is ample space for carrying out cleaning and maintenance. There must be adequate space to remove the flux and solder pots, to clean off the dross from the solder surface, and from time to time both the flux and solder pots will require draining and cleaning. For the solder pot, in particular, this requires a clear area, since the molten solder will have to be drained into suitable molds, and the process of cleaning out the pot may require the use of tools, brushes, and so on. Refilling the pot can also be carried out more efficiently by melting the solder in a separate container, and this too requires a free area around the machine.

The actual work area itself should be clean, bright, and well lighted. One of the objectives is to maintain an excellent level of cleanliness and housekeeping in the soldering process and this will not happen unless the area is conducive to this method of working. In the same vein there must be adequate storage space for the supplies that are necessary for the day-to-day operation and maintenance of the process: shelves, cupboards, and a bench adjacent to the machine, not forgetting that some of the fluxes and solvent may be flammable and require fireproof storage containers or a fireproof cabinet with external venting if the amount stored is more than a nominal quantity. A sealed container must also be provided for dross, and ideally there must be washing facilities so that the operators can wash their hands immediately after handling solder.

The flux pot will have to be dumped from time to time, and there must be some arrangement made for disposal of the spent flux and the washings when the pot is cleaned out with a solvent. Depending on the flux used, this may be considered as hazardous waste and require special storage and disposal arrangements. Check with your safety officer or state authorities.

The regular solder analysis will call for a ladle and molds into which the samples are allowed to solidify and cool. These items should be used for this purpose only, and storage space for this equipment must be arranged. Similarly, the importance of recording all the process details has been emphasized again and again, and a clean area must be maintained for this purpose only. It should have a desk and a bulletin board where the various charts and graphs can be prominently displayed. Only in this way will the records be maintained accurately and used effectively.

The smaller machines are not difficult to move if for any reason the circumstances alter, and another plant layout is necessary. The larger systems, however, especially where the solder machine is part of a total assembly, cleaning, and test line, present a major task if a change in layout is necessary. Therefore careful planning and consideration of the previous comments are most important. Remember all machines require venting to the outside air, some will need a compressed air supply, and of course all need a power supply of the correct voltage and capacity. Although there has been no report of any damage to components during soldering through static charges or leakage currents, it is theoretically possible for such potentials to exist, and the best way to prevent any problems is to ensure that the machine is correctly grounded.

The compressed air supply is usually needed for air knives and fluxer, although two manufacturers use compressed air motors to drive the solder pump. For the air knives and fluxer it is absolutely necessary that the air be completely dry, clean, and especially free from any oil. If the machine vendor does not supply an air cleaner, this should be added during installation. For operating air motors, the manufacturer's instructions should be followed regarding air requirements, including filtering and lubricating.

Within the soldering area the safety notices must be prominently posted giving clear details of the safety precautions to be followed. These rules must be clear, precise, and rigidly adhered to.

With the machine on site and all facilities connected, the first task is to make an initial check of the system and the machine setup. Note that this is a check, not a calibration or the setting of soldering parameters; these come later. All machines are fitted with leveling bolts, and the first action must be to see that the conveyor is level from front to back and that the solder wave nozzle lip is also perfectly horizontal. This is to ensure an even contact between the solder wave and the board, over the entire width of the wave.

The manufacturer generally supplies instructions for doing this and these must be followed precisely. Now the supply to the machine can be switched on.

With the solder pot heaters off, try out the operation of the rest of the machine. Run the conveyor, look for a smooth, quite operation with no jerking, or speed changes. Operate the speed control and see that it varies the conveyor over a reasonable range. Switch on the preheaters, and see that they heat up. Fill the fluxer with clean flux, and check that a wave or foam head of the correct height is obtainable, and that all the controls function. If a flux density controller is fitted, make sure that the pumps operate correctly, and that the density sensor is functioning.

Once it has been determined that there is no major problem it is time to turn on the solder pot. The reason for leaving this until last is simply that the machine is much more difficult and inconvenient to work on once the pot is full and hot. If possible melt the solder in a suitable container and pour it into the solder pot. This is the more convenient way to go, and almost any stainless steel container is suitable and any heat source can be used. The pot, of course, must be clean and not contaminate the solder in any way. If this is not possible, then the solder pot will have to be charged with solder chips or pellets. This is necessary to prevent burning out the heaters by running them in an unloaded condition. Unless the greater part of the heat generated by the heaters is removed by conduction into the solder, hot spots will be formed which may cause permanent damage. In some machines it may be necessary to disconnect some of the heaters until a certain level of molten solder is obtained. This will vary from machine to machine, and the manufacturer's instructions should always be consulted.

Once the solder pot is fully charged and the solder is up to operating temperature some initial checks of the accuracy of the various subsystems should be made. Check the speed of the conveyor at several settings of the controller. Do not expect the highest accuracy with the simpler systems. Remember that repeatability is much more important than absolute accuracy. The speed can easily be measured as described in the section on conveyors.

With a mercury in glass thermometer make a check of the solder temperature. To avoid errors, the thermometer should be placed alongside the controlling thermocouple. If this is not possible, there may be some small temperature difference until the wave has been run for some time to stabilize the system. Once again, a small error in the controller may be acceptable provided that the difference is known and can be compensated for by changing the controller set point. Most temperature controllers can be quite easily recalibrated, although some manufacturers are reluctant to disclose the very simple procedure, and changing the calibration may negate the warranty on the instrument. Here again repeatability is one of the more important factors.

If the preheaters have full temperature control with thermocouple and controller, they should be checked in the same way. It may be difficult to make this check with a mercury in glass thermometer, and a calibrated thermocouple and readout will make the task easier. With the hot plate type heaters some variation in temperature can be expected across the surface. This is not of major concern, because the spacing between heater and board will tend to reduce the actual variation as seen by the assembly. If the variation exceeds 20°F (12°C), it may indicate a failed heater. Once again the actual temperature is not as important as the repeatability of the controller setting. If the appliance type of controls are fitted, then tests should be made to ensure that they are working correctly, and some measure of their repeatability obtained, together with their overall control range.

The wave should now be run and tested for height, levelness, and smoothness, and some experience gained in setting up and operating the wave controls. Adjustments and calibration of the solder wave can be accurately carried out only by using a heat resistant glass plate (Fig. 5.80). The plate is

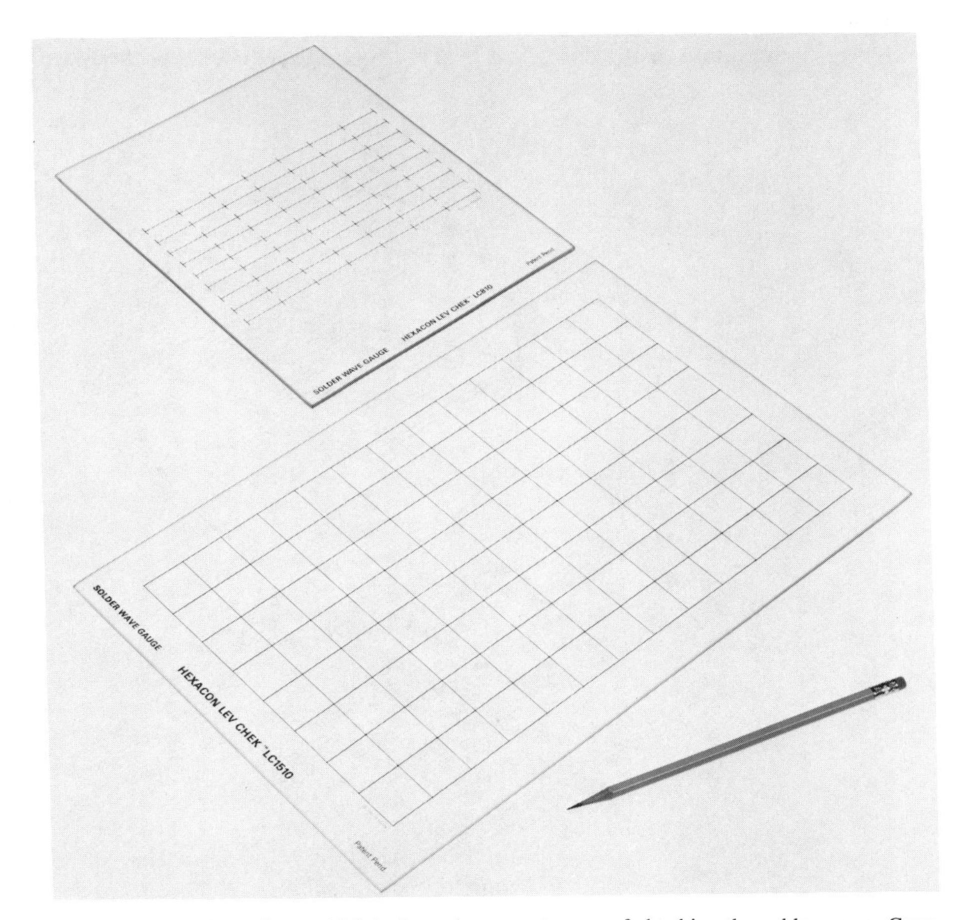

Figure 5.80. A glass plate, which is the only accurate way of checking the solder wave. Courtesy Hexacon Electric Co.

run over the machine in place of the board assembly. It is fluxed, preheated, and then run slowly over the solder wave. It can be stopped in contact with the wave for a short time without cracking (Fig. 5.81). The area of solder in contact with the plate will be clearly visible, and any necessary adjustments made to obtain an even coverage. The solder nozzle and the machine conveyor having been correctly leveled, there should be little to change, but this is a quick and simple test, which should be part of the daily machine check. Tempered or other heat resistant glass is ideal for the purpose, or a sheet of epoxy glass can be used as a temporary substitute. There are specially made plates on the market designed for this purpose. With the help of these plates it is also possible to check the amount of flux applied to the assembly, and even some assesment of the adequacy of the preheat can be made. As the plate passes over the fluxer, the width of fluxing can be checked, and any disturbances in the foam or wave noted and rectified. By the time the plate reaches the solder wave, much of the solvent in the flux will have been evaporated, and any boiling of the flux or solder spitting will indicate that the preheat is inadequate or the conveyor is set at too high a speed. As the plate

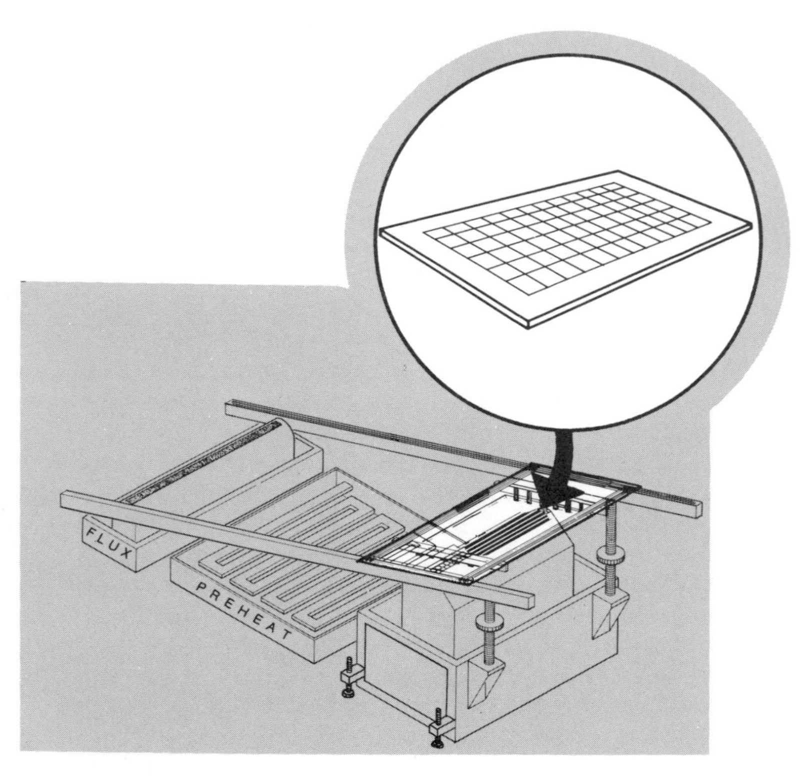

Figure 5.81. This diagram illustrates the glass plate being used to set the solder wave height. Courtesy Hexacon Electric Co.

exits the wave there should still be a thin line of flux visible, at least 1/32 in. (1 mm). If this is absent, it shows that either too little flux was applied or the solids content was too low.

With the plate stopped over the wave the entire contour of the solder surface can be seen and the contact lime calculated by multiplying the thickness of the wave, as measured on the plate, by 60 (seconds per minute), and dividing by the conveyor speed in inches per minute.

Oil versus No Oil

There has been a great deal of controversy during the past few years on the use of oil in the solder wave of the mass soldering process. This stems from the fact that of two major manufacturers of wave soldering equipment, one produces machines with oil injection in the wave, and advocates this as the best method of achieving high-quality joints, while the other is a strong supporter of the advantages of the adjustable asymmetrical wave, used without oils, the so-called *dry wave*.

Both companies have their proponents, who have publicized their own preferences very much in isolation of the other. Each company holds patents on the process that they recommend. It is not easy, therefore, for the newcomer to soldering to differentiate between fact and publicity and between genuine experimental results and subjective testing. It is not that there is any attempt to deliberately distort the facts, but when emotion comes in the door, objectivity goes out of the window. This is an attempt to lay down the facts as honestly as possible, although the author was involved with one of these companies.

Oil injection is the much older system. As discussed in the section on nozzles the two main problems with the symmetrical wave were the small contact area with the board, which restricted the speed of soldering, and the tendency to produce shorts and icicles. The latter problem was tied in with the speed of the falling wave at the exit point of the board from the solder, and the angle between the board and the wave at this point. These factors resulted in the inability of the excess solder to drain from the joints, especially as the soldering speed was increased.

At that time the idea was proposed of achieving the required solder drainage by reducing the surface tension of the solder. There were several methods evolved of using oil to do this, and one of the most elegant was to pump the oil through an adjustable nozzle into the solder at or near the pump impeller.

This produced a mass of tiny globules, which were carried to the surface of the wave, where they burst and covered it with a thin layer of oil, while some remained intermixed with the solder in the form of finely dispersed particles. Together they effectively reduced the surface tension and encouraged the solder to drain off the board at the exit point from the wave, leaving only the minimum necessary to provide a satisfactory electrical and mechanical connection.

The system works, and works extremely well—dramatically reducing the number of solder joint failures produced by the symmetrical wave. The oil ends up on the surface of the solder pot and helps to reduce the formation of dross by excluding the air from the solder surface. It has also been suggested that the oil will assist in maintaining the solder purity by combining with some of the metallic contaminants, which are then removed with the excess oil, although the author has never seen any proof of this.

These advantages are not achieved without some additional problems and expense:

The oil is not cheap, and the cost has to be added to the cost of the soldering process, although the reduction in the amount of dross produced will compensate to some degree.

The oil that drains off the pot must eventually be disposed of; environmental regulations may require special treatment at additional cost.

Water soluble oils must be used if aqueous fluxes and cleaning are part of the process.

Cleaning has to remove the oil as well as the flux.

The oil injection system requires regular maintenance and cleaning.

The oil can so reduce the surface tension that empty plated through holes, and holes with a large ratio of lead to hole diameter, will sometimes drain completely free of solder leaving a joint that is probably sound, but contravenes some specifications.

When the asymetrical wave was developed, higher soldering speeds were possible, and the use of oil in the wave became even more important to achieving good results.

Eventually the adjustable asymmetrical wave was developed, in which the excess solder was removed by the intrinsic shape of the wave and the flow pattern of the solder. This achieved similar results to those obtained from the previous waves but without the problems associated with the use of oil. The new wave shape proved to be extremely effective and indeed has become an industry standard for fast, high-quality soldering, achieving solder drainage equal to that obtained by the use of oil. As with all processes the use of this wave has some negative features:

Once set up, a major change in wave height requires readjustment of the nozzle and jacking of the solder pot to compensate.

There is no oil in the surface of the wave or the solder, and the rate of dross production is only limited by the design of the pot and nozzle.

The wave must be adjusted carefully to produce optimum results.

This wave shape can only be used with an inclined conveyor.

The question of oil or no oil, therefore, is not as simple as may appear at first glance, and any decision on this question should be based on the overall requirement of the system. Review the various sections of any machine and decide that together they will provide the necessary capability. Ask the vendors of the selected systems to give you the opportunity to carry out tests on a demonstration machine. No reputable manufacturer will refuse this request. The following suggestions will help to make the tests truly representative:

Choose the most complex board to be processed.

Run enough boards to give at least 5000 joints for inspection.

Solder one batch of boards with oil, on a machine specifically designed for this process.

Solder a second batch on a machine designed solely to use a wave without oil. Do not carry out these tests by using the same machine and just shut off the oil; optimum soldering without oil is only obtained with nozzles specifically designed for this purpose.

In both cases obtain cost figures for the usage of solder, flux, and oil.

Use an independent inspector to check the results after cleaning the boards.

Compare the results: number of reject joints, any shorts or icicles, and if throughput is important, what was the maximum soldering speed? Estimate the cost of running the different systems. Evaluate the entire operation for each machine type. Assess maintenance costs and possible down time, the effects on the cost of cleaning, capital expenditure, and so on.

But above all, look at the soldering results; bad joints are the most expensive part of any soldering system. Let the results be your guide, not the publicity or the opinions of others.

Chapter 6

Setting Up and Operating

Once the machine has been purchased and installed, the most interesting part of the setting-up process can begin. At this time all the information gleaned from the earlier chapters of this book will come into play, and the advantages of the testing and comparing of the various systems will be obvious.

From the testing and the installation work already completed, the machine should feel familiar and the actions of the various controls well understood. Do not rush the task of setting the operating parameters. This is an extremely important part of the operation and will be the basis of the soldering process for a long time to come. Make absolutely sure that all of the checks and calibrations have been properly done and recorded in case some change is inadvertently made to the machine, and it becomes necessary to go back to square one. Set out a record sheet before touching the machine, and once the work begins allow only one person to make any adjustments. It is convenient to have a second individual to observe and record, and in this way no inadvertent changes will be made.

SETTING PARAMETERS

For every different design of board it will be necessary to develop the correct soldering parameters. For this purpose it will be necessary to have a supply of assemblies, at least 20 of each type. With experience it will be easier to determine the correct settings, and fewer boards will be required, but do not be tempted to accept the results from a few boards as being typical of the lot. A reasonable sample size is needed to ensure consistent results during production.

Start with the most complex board to be processed. This may appear to be going the wrong way, but the chances are that having determined the settings for this assembly, they can be applied almost without change to the less difficult boards. Complexity of design means an assembly with tightly packaged components and closely spaced conductors and mounting pads.

Most of the soldering parameters are interdependent, and therefore some initial setting must be made in order to begin testing. It is generally agreed that preheat should be between 200°F (95°C) and 260°F (127°C) when measured on the top surface of the board immediately before the solder wave, and this is a good starting point. Set the conveyor to the soldering speed at which it is required to run in order to obtain the necessary output. If this is not an important factor, then start with a speed of 6 ft/min (1.8 m/min). Turn off the fluxer and the solder wave, and set the preheater to a midposition. Now run an assembly and measure the temperature of the top surface of the board just as it reaches the soldering station. There are several ways to do this, but one of the simplest is to use the temperature indicating crayons or labels. The former melt at a predetermined temperature, while the labels change color. For example, if a 200°F (95°C) board temperature is required, then a 200°F (95°C) crayon or sticker will be used, and one each 20°F (11°C) above and below this temperature. As the boards pass over the preheaters, the lower temperature crayon mark will be seen to melt, or the label to change color, then the required temperature indicators will change, while the highest temperature ones will not change at all. Of course a thermocouple and suitable readout can also be used, although it is not easy to couple to the board, and the question of response time then enters the picture. There are some special temperature measurement systems available, but unless very accurate testing is to be carried out the crayons and labels offer the most practical temperature indicating system and are accurate enough for most setup purposes (Fig. 6.1).

Figure 6.1. Some typical temperature indicating wax crayons and color change labels.

It is unlikely that the correct temperature will be obtained on the first run, so continue to make small adjustments to the preheater controls until the required temperature is obtained. Remember that the preheaters will take some time to change temperature, and the assembly must be allowed to cool between tests. Once the correct setting of the preheater controls has been established, and the same top of the board temperature can be obtained consistently on a few assemblies, it is time to actually start fluxing and soldering some product.

Put on the fluxer and run the assembly over the fluxer and the preheat. The underside of the board should be examined when it exits the machine. The flux should have coated the bottom of the board evenly, and there should be evidence of the flux passing up through any via holes to the top of the board. When the bottom of the board is touched lightly the flux should feel tacky or sticky, but should not be wet. If it has not dried sufficiently, check the operation of the fluxer and be sure that the fluxer air knife or brush is removing the excess flux. In case of difficulty in achieving this result, check with the flux vendor, and follow the vendor's recommendations. There are so many variations in flux composition that it is not possible to give more than very general instructions. With some of the water soluble fluxes in particular it may not be possible to achieve the same sticky consistency as with the rosin fluxes.

When several passes have to be made to arrive at the correct setting, remember to use a different assembly each time and allow them to cool down between tests. The board temperature when placed over the fluxer can have quite an effect on the amount of flux deposited, especially with the foam system.

Now that the fluxing and preheat have been set up, the wave should be switched on and adjusted to the correct setting for the wave shape and wave height. If oil is used in the wave, set the oil valve in a minimum position. For safety, always wear approved glasses or goggles when operating the machine with the wave exposed and follow the safety rules. Now run an assembly through the entire process, and when cool, clean off the flux and flux residues and examine the results. After a review with the naked eye, use a magnifier, up to 10 power, to check any doubtful joints. Note any failures and then run a second assembly. Carry out the same inspection and see if any failures are repeated, or if random failures are seen. It may require several test pieces to be run before a real determination can be made. The random failures should be ignored at this time. They are probably due to solderability problems rather than the setup of the machine. Repeated failures will indicate incorrect adjustment of some of the solder wave-parameters. The machine manufacturer's instructions should be followed carefully until satisfactory results are obtained.

There are so many different types of solder machines that to give detailed instructions is quite impossible; however, the information in the previous chapters should have provided sufficient knowledge to enable any soldering system to be tackled in a logical manner and optimum adjustments achieved once a little experience is gained in the operation of the controls.

There are some general rules, however, that will be found useful in arriving at the correct machine setting:

Make sure that all the boards and components have passed an approved solderability test. The problem of setting the machine correctly are multiplied if the solderability of the parts is in question.

Do not change assembly types until one set of parameters has been established.

Run sufficient assemblies at one setting to assure that any randomness is not confusing the results.

Never change more than one parameter at a time, and only make small changes.

Remember that any change in conveyor speed will affect both preheat and immersion time in the solder. The soldering time is the most critical. Once this has been established preheat can be adjusted to give the correct board temperature.

Once an acceptable preheat temperature has been established (for example, the maximum that will not damage either the board or the components), then setting the machine consists of adjusting the conveyor to the highest speed at which the most thermally massive joints in the assembly are completely wetted, while maintaining the accepted preheat temperature.

As soon as the parameters for a particular assembly have been established they must be incorporated into the process document for that assembly and no deviation permitted under any circumstances. This cannot be too strongly emphasized, and it must be part of the QC function to see that this rule is rigidly enforced.

With changes in board type, for example, from single-sided to double-sided with plated through holes, there will be considerable changes to the processing parameters. Because of the need to bring the entire mass of the through hole plating up to soldering temperature, either more preheat or a slower conveyor speed may be necessary. Obviously, if it is necessary to maintain a certain level of output, then the conveyor speed cannot be reduced, and the entire additional heating must come from the preheaters. The number and type of components will also affect the heating requirements, as will variations in the leads or pins that form part of the joint. With plated through holes the specifications usually require that a top fillet be established to prove the soundness of the solder joint, and this requires that the top of the PWB and the adjoining circuitry are also brought up to soldering temperature.

Multilayer boards require even more preheat, depending on the number

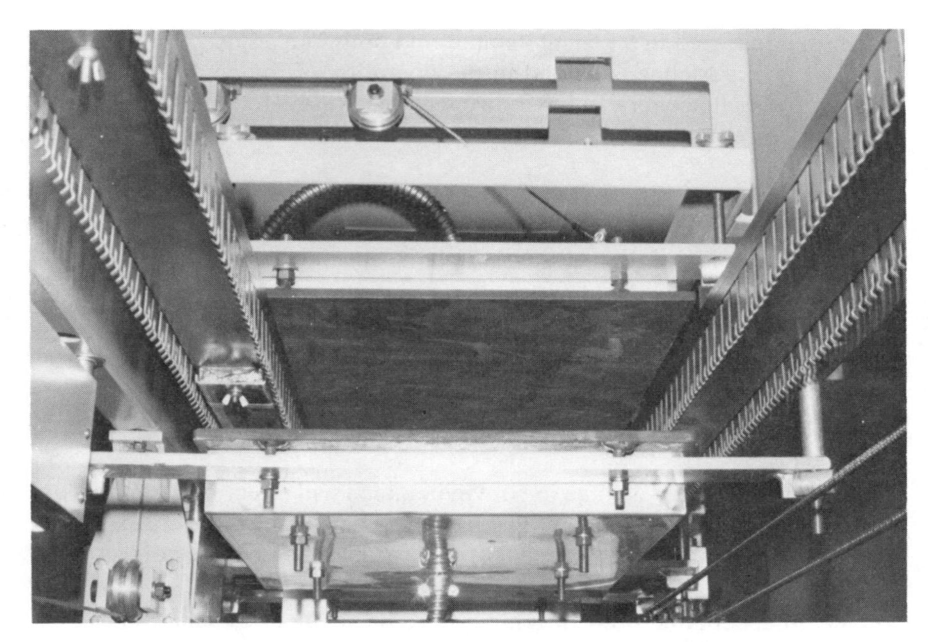

Figure 6.2. A top preheater. Courtesy Electrovert Ltd.

and thickness of the internal layers and the area of copper that they contain. When thick internal layers are used, their thermal capacity can become a major problem, and it may prove difficult to provide sufficient preheat to retain the higher soldering speeds. Boards with many layers have the same problem, simply because of the mass of copper that has to be raised to soldering temperature. For these multilayer boards, top preheat becomes a necessity, unless extremely slow soldering speeds are acceptable. Of course, care has to be taken to avoid damage to the components, and this is usually the limiting factor when top preheat is used. Fortunately components are being steadily improved in this area, and in consequence preheat temperatures are being increased, with a corresponding increase in soldering speeds (Fig. 6.2).

CONTROLLING THE PROCESS

Soldering is a process and like any other process will only perform correctly and consistently if the operating conditions are carefully controlled and any changes made with care, in a logical and cautious manner.

> *Soldering is easy. All you need are the correct flux and solder, the right temperature for the right time, and solderable parts.*

Unfortunately, it is not always easy to provide these items, because of the many variables involved. Therefore, if the process is to have any chance at all to produce consistent results, everything possible must be done to remove these variables from the operation.

The importance of these items to the overall process has been discussed in detail in earlier chapters; now the methods of setting up a formal control procedure must be defined. Process instructions must be prepared for all of the machine functions. These will vary from machine to machine, depending on the complexity of the system and the layout of the various stations. There are, however, certain functions in all soldering machines, and examples will be given of these, as follows.

Setting up and operating the fluxing system
Operating instructions for the preheaters
Setting up and running the soldering station
Defining the machine operating parameters

There may well be other functions, such as the finger cleaner, automatic density controller, and so on. If the examples given are understood, then there should be no difficulty in developing the required documentation for these other special processes.

Control Document–Fluxing Station (Fig. 6.3)

This process document must contain all the instructions necessary for the operator and QC inspector to set up and control this process. As a minimum it must include the following:

Type of fluxer, make, and serial number (where more than one is used in the plant)
Instructions for setting up, wave or foam height, control settings, such as air pressure, fluxer height, air knife angle, and so on
Type of flux to be used, vendor's name, any data required for identification, such as color, density, smell, and so on; fluxes are occasionally incorrectly marked
Method of carrying out density measurements, frequency of testing, and a table of the amount of thinners to be added for a range of density readings
Frequency of flux dumping and the quantity to be added to refill the fluxer
Method of disposal of the spent flux
Detailed instructions on record keeping (Fig. 6.4)
Frequency of maintenance, referencing the appropriate maintenance document
QC inspections to be made on the process

CONTROL DOCUMENT FLUXING. NUMBER 1234.

FLUXER TYPE. Foam fluxer 605A. Ser: No. 497.

FLUX TO BE USED. Alka foaming flux type 697.

DESCRIPTION. This flux is a light blue, and has a strong smell of alcohol. The solids content is 30% and it has a density of 0.89. The flux container must be kept closed at all times to prevent evaporation. Only Alka thinners type 142 to be used. NOTE. All these materials are flammable.

Fluxer Set Up. The top of the nozzle shall be 1/4 inch 6mm from the bottom of the board.
Fill the fluxer with flux 697 to the level mark.
Fill the feeder bottle and invert into the holder.
Open the air valve until the gage reads 12psi.
Slowly open the control valve until a full head of foam emerges from the nozzle.
Continue to open the valve until the foam is broken by large bubbles.
Very slowly close the control valve until the large bubbles stop. --- Q.C. to monitor.---

Density Control. Check the flux density once every hour of operation. Use hydrometer part No. 285. Record the reading, together with any flux or thinner additions in the daily record. Add Alka thinners 142 according to the table below.
Note. The density readings must not be made until the fluxer reaches a steady temperature. This requires 30 minutes of machine running. --Q.C. to monitor.--

Thinner Additions.

Density readings.	--0.891	0.892	0.893	0.894	0.895
Add litres of 142	--0.28	0.30	0.32	0.34	0.36

Flux Additions. When the reservoir bottle is less than 1/4 full it must be refilled with 697 flux.

Shut Down. If the fluxer is to be shut down for more than one hour, place the stone in the storage container. This must always be kept full of 142 thinners, and the lid closed.

Cleaning and Maintenance. At the end of each shift the fluxer must be cleaned with 142 thinners, on all external surfaces, and the air knife removed and washed out by brushing with thinners, followed by a rinse in warm water. After drying replace, being careful to retain the 5deg angle according to the calibration marks. **Every Saturday** the fluxer shall be removed and the flux drained. The fluxer shall be cleaned and maintenance performed according to maintenance document 1248. -- Q.C. to monitor.---

Figure 6.3. A simulated control document for fluxing.

Control Document—Preheating (Fig. 6.5)

Make, type, and serial number of the preheater (where more than one is used in the plant)

Setting up on the machine: height from the conveyor, and so on

Time to heat up from cold and time to be allowed between making control changes and running product

Detailed instructions on record keeping (Fig. 6.4)

MACHINE SOLDERING STATION NUMBER 2 DAILY RECORD. DATE 6-3-82 SHEET No. 1

Assy No	Ser No	Flux data	Pre-heat	Solder	Conv	Time	Oper	Q.C.Sig.
48501	820 – 830	Alka 697 0·89	7-7-4-4	485°F	6·8	7·15	T.J.	
		0·89				7·30	TJ.	
J6823	1010 – 1110	Alka 697 0·89	7-6-4-5	485°F	7·2	7·45	TJ.	7·55 TS.
		0·891 – Added 0·28L Thinners				8·30	TJ.	
		0·89		Added 1 bar		9·00	TJ.	
	Completed					9·05	TJ.	
J6824	068–168	Alka 697	6-6-7-7	485°F	5·6	9·15	TJ.	
		0·89				9·30	TJ.	
	Completed	0·89				10·00	TJ.	10·15 PS
48502	880 –890	Alka 697	9-9-9-9	485°F	4·5	10·30	T.J.	
		0·891				11·00		
	Completed	Added 0·28l Thinners				11·15	TJ.	

Figure 6.4. A simulated daily record of product passing through the solder machine.

Frequency of maintenance, referencing the appropriate maintenance document

QC inspections to be made on the process

It may appear that this degree of control is unnecessary with something as simple as a hot plate preheater, but consider that the plate may be removed for repair or cleaning and replaced at a different distance from the conveyor. The resulting change in the board temperature may not be noticed for some time during which the incidence of failed joints could have increased beyond that acceptable in zero defect soldering. Similarly, unless regular QC checks of the operating temperature are made, the heater could suffer a failed element and the operator be unaware until the soldering results made an investigation necessary.

Control Document—Solder Station (Fig. 6.6)

In a similar manner to the fluxer control document, the process instructions must cover all the functions necessary to set up and operate the station, and as a minimum include the following:

Type, make, and serial number of the pot and nozzle (where more than one is used in the plant)

Composition of the solder to be used, suppliers and specification to which purchased

Frequency of solder analysis, instructions for taking samples, methods of recording results

Setup procedure for nozzle; frequency of check of adjustment

Frequency of dross removal; instructions on method to be followed and disposal

Frequency and method of adding fresh solder

CONTROL DOCUMENT PRE-HEAT. NUMBER 3455.

PRE-HEATER TYPE. Hot Air 640. Hot Plate 85.

Set Up. With a board mounted on the conveyor the hot air 640 must be 1.5 inches, 37mm, from the bottom of the board to the top edge of the louvres.
The hot plate 85 must be mounted with 1inch,25mm, clearance from the bottom of the board.
Both pre-heaters must be centralised under the conveyor. The hot air 640 must be 4inches, 100mm, from the corner of the cover to the edge of the fluxer tray. The hot plate 85 must be 4inches, 100mm from the cover of the hot air 640.

Operation. From switch on 45 minutes must be allowed for the pre-heaters to reach operating temperature. During this time no product must be run, and the machine doors must be closed.
Temperature changes around the operating temperature will take about one minute for every degree of change. These times must be noted and no product run until sufficient time has elapsed for the pre-heaters to reach the new temperature.
Note that the operation of the controllers cannot be used as an indication that the heaters are up to temperature.

Recording. Every time that any change is made to the pre-heat temperature it must be logged in the daily report, with the time that the change was made,the time allowed for the pre-heaters to reach the new temperature, and the time that product was again run.

Maintenance and Cleaning. The hot air 640 must have all flux wiped off at the end of every shift. Every Saturday the filter must be removed and washed out according to the manufacturers instructions. The hot plate 85 requires no cleaning and maintenance must be carried out according to maintenance document 1248

Q.C. Requirements. The position of the fluxers to be checked once each week, after cleaning and any maintenance. On a random basis temperatures must be checked to the Operating Parameters.
Regular checks of maintenance to be made according to the normal Q.C. schedule.

Figure 6.5. A simulated control document for preheating.

CONTROL DOCUMENT SOLDER STATION. NUMBER 386.

SOLDER POT. Type HTS 450. Ser; No. 285.

SOLDER TO BE USED. 60/40 alloy to QQS 571E.

NOZZLE. Assymetrical type HS by Solder-Rite.

Set up. The thermostat for the solder pot must be set at 482F 250C. It will take approximately 4hrs for the solder to melt and reach operating temperature. Switch on the solder pump, and adjust the speed to give 1/4inch depth of solder over the top of the nozzle lip,(6mm). Slacken off the screws holding the back plate of the nozzle, and adjust until there is no solder flow over the plate. Now increase the solder height to 5/8inch, when the solder should just trickle over the back of the nozzle (15mm). If not readjust the plate. This set up is to be carried out at the start of each shift.
-- Q.C. to monitor. --

Solder Analysis. The solder is to be analysed at the end of each month or as requested by the S.P.C. The sample is to be taken at the end of the shift, from the wave, using the ladle provided in the kit. The solder is to be poured into the disposable dish, and allowed to cool. It is then to be scribed with the date and the next serial number from the solder analysis book. It is then to be sent to one of the approved solder vendors for analysis, via purchasing. Results are to be plotted on the control chart, and then sent to the S.P.C. Any out of tolerance condition is to be reported immediately to the S.P.C.

Dross Removal. At the start of each shift the dross is to be scraped from the pot surface using only the tools provided.A mask and gloves must be worn, and the dross removed placed immediately into the dross container. When full purchasing is to be advised, and they will arrange for removal. -- Q.c. to monitor.--

Solder Addition. When the level of the solder in the pot falls below 1inch, 25mm from the pot edge, fresh solder must be added. Four one pound sticks must be laid, one at a time in the molten solder beside the nozzle. Allow each stick to melt before adding another.

Solder dumping. Whenever the solder analysis indicates an out of tolerance condition, or every six months, which ever is the shorter, the solder will be drained from the pot, the pot cleaned of all dross and oxide, and refilled with fresh solder. 450 lbs,205 kgs is required. The drained solder is to be returned to the vendor for credit.

Maintenance and Cleaning. At the end of each shift the pot is to be checked for dross and solder level as already described. the moter, belt and drive is to be checked for wear, and smooth running. Any solder splashes are to be removed, and accumulations of dross or oxide removed from the nozzle. If the pot is to be shut down, it must be covered with the heatproof fabric cover provided, and the time clock checked to assure that the heaters will be switched on at the proper time for the next shift.
Every Saturday complete maintenance is to be carried out according to the maintenance document 1248.
--- Q.C. to monitor. ---

Figure 6.6. A simulated control document for the solder station.

Page 2 of 2.

Special Note. Because of the importance of maintaining the solder purity nothing must be placed in the solder except the special stainless steel tools provided. Any accidental contact with the solder of any other material must be reported immediately to the S.P.C. --- Q.C. is to monitor the solder purity most carefully, together with the analysis records.----

Safety. Any person handling solder, is to be sure and wash their hands before eating or drinking to avoid ingesting lead. The safety notices must be followed completely, and approved masks worn when removing dross from the pot. FOLLOW THE SAFETY RULES.

Figure 6.6. A simulated control document for the solder station.

Detailed instructions on record keeping (Fig. 6.4)

Frequency of dumping solder, instructions on the procedure to be followed, quantity of fresh solder required to refill the pot

Frequency of maintenance, referencing the appropriate maintenance document

QC inspections to be made on the process

Control Document–Oil Injection

Where oil is used in the wave, it becomes an integral part of the soldering process, and must be controlled as strictly as any other function. Because it is physically part of the solder pot, the system should be included in the solder documentation. It is separated here because only certain machines use oil in the solder.

Make, type, and serial number of the system (where more than one system is used in the plant)

Manufacturer and identification of the oil to be used

Instructions for setting the nozzle and filling with oil

Instructions for draining the used oil from the pot, frequency, method to be used, and instructions for disposal of the spent oil

Detailed instructions on record keeping

Frequency of maintenance, referencing the appropriate maintenance document

QC inspections to be made on the process

These basic documents will control the setting up of the various parts of the machine, the operation, and the maintenance, together with the QC overview to see that everything is carried out according to the instructions. These documents will apply to all the products processed on the machine. It is unlikly that product variations will be so great as to require changes to these parameters. It is just possible that the nozzle may have to be changed for some particular assembly, for example, one with very long leads, and in this case an additional document will have to be prepared describing the setup of the soldering station with this different nozzle. Similarly, other changes may occasionally be made, but the likelihood is remote, and sufficient examples of the documentation and control required have been given for the correct instructions to be developed.

However, for every different design of assembly to be soldered a document must be generated giving the particular settings of the machine necessary to process that board correctly. Even if the same settings can be used for more than one assembly, it is much safer and more convenient to set out a separate setup document for each one, using the type number or drawing number of the assembly as the identification for the processing document. Fortunately, the data required are not many, because all the standard parts of the machine setup and control are contained in the control documents just discussed.

SOLDER MACHINE OPERATING PARAMETERS

The document that will determine the actual day-to-day operation of the machine contains the operating parameters. Having been set up and checked according to the control documents, the operator will now look at the part number or drawing number of the parts to be soldered, and from a library of instructions pull out the relevant data to make the final adjustments before actually starting to run the product. These data therefore must reference some of the control documents, especially where different fluxers, nozzles, or other systems are used on the same machine, so that the operator can be sure that the correct systems are installed.

He will then set the various controls on the machine to the operating parameters, and the process can begin. Note that there is no trial and error in the start-up and no experimental settings.

A typical set of machine operating parameters therefore will include as a minimum the following items. Because this is an important document a sample is shown in Fig. 6.7.

Serial number, type number, or other identification of the relevant assembly

Fluxer control document to be used

Solder control document to be used

Preheater control document to be used

Any changes to these documents for this particular assembly

Settings of the preheater temperature controls, or preheater temperature, depending on the preheater control system

Solder pot temperature

Conveyor speed

Conveyor angle (if a variable angle conveyor is used)

Any other particular instructions regarding the soldering of the assembly, for example, it may be necessary to orient the assembly in a particular way to optimize soldering

To summarize this control system, the philosophy is to put as much of the common instructions into the control documents for the various parts of the machine. Those instructions that vary with the assembly to be soldered are placed in the solder machine operating parameters. The details given here

```
OPERATING PARAMETERS.  SOLDER MACHINE.  No: 6542.

MACHINE TYPE. Solder-best 21.        SER No: 1458.

REFERENCED DOCs. Fluxer 3456. Pre-heat 1234.

                 Solder station 987. Maint: 4536.

THESE OPERATING PARAMETERS TO BE USED FOR THE FOLLOWING

ASSEMBLIES.      9876534    9876532  9876540

                *658934   *658935

   CONVEYOR SPEED. 6.8 ft per minute.

   FLUXER AIR KNIFE.   5degrees  0.8 psi.

   SOLDER TEMP.        485F.

   SOLDER WAVE. Set at 3/8inch, 10mm.

   POT HEIGHT.  Set to give an immersion depth
                of half of the board thickness
                or a solder contact of 2.4inches
                62mm, when measured on glass
                plate part no, 4567 placed in
                finger conveyor per doc.987.

   PRE HEAT.    Heater #1 control at setting 8.
                Heater #2 control at setting 6.
                Heater #3 control at setting 6.
                Heater #4 control at setting 7.

   CONVEYOR WIDTH. Set with template No. 9981.

   NOTE. Assemblies marked * must be run with the
         connector leading. The remainder must
         have the fingers taped and placed with
         them turned to the back of the machine.

   RECORDS. All assemblies soldered must be entered
            into the daily record, referencing this
            document.
```

Figure 6.7. A simulated control document for the operator to use in setting up the solder machine.

are applicable to most installations, but of course must be changed where local requirements make an alternative format simpler to use. For example, where one solder machine only is in use and only one set of operating parameters are necessary, there is no reason why the entire instructions should not be put together into one document. The essential thing is that no part of the process, no matter how small and apparently insignificant, must be left out of the control loop.

The aim is zero defect soldering, and there is no room for variations from the optimum processing parameters. When a problem is discovered the solution will be found by logical thinking, and accurate statistical records, not by tweaking the settings of the solder machine.

MAINTENANCE

Maintenance is as much a part of the soldering process as the correct setting-up of the machine, and must be controlled in the same way. Working from the manufacturer's recommendations, a schedule of maintenance and cleaning must be established and, most important, once established, it must be rigidly adhered to. There is always a tendency to put off this work for what at the time appears to be a very good reason. The machine is working fine, production is behind schedule, and so on. This must not be allowed to happen. It is only with good regular maintenance that the machine will function consistently and possible variations in machine performance avoided. The maintenance document must describe all the work to be carried out on the machine, define who is responsible for doing it, and the QC function to check that it is carried out correctly and to schedule. It must include the following:

Daily cleanup at the end of each shift

A weekly clean with a general check of performance

A thorough strip down and clean at about once a month; lubrication of moving parts, general check, and readjustment

Approximately every 6 months or as the solder analysis requires, if sooner, dump the solder, clean out the pot

The manufacturer's recommendations must be followed in preparing these instructions. A typical schedule is shown in Fig. 6.8.

QUALITY CONTROL OF THE SOLDER MACHINE

The final control of the machine performance must be part of the QC function. As has already been pointed out, all the control instructions for the machine must come under the scrutiny of QC to ensure that they are carried

CONTROL DOCUMENT. MAINTENANCE NUMBER **1248.**

SOLDERING MACHINE. Solder-best 21.
Pre-heater hot air 640 hot plate 85.
Foam fluxer 697 Solder pot HTS 450.

Daily Clean. The entire machine is to be cleaned at the end of each shift, of any flux or solder spills. Dross to be removed, and the surface wiped down with a cloth dipped in a 5% solution of hot detergent. The individual modules are to be cleaned according to the appropriate control documents.

Weekly Maintenance. Every week end, Saturday unless work schedules make some other time more convenient, the machine is to be completely cleaned, the fluxing module removed, drained, washed out, and replaced, the stone checked for cleanliness and porosity according to the manufacturers instructions.
The pre-heater is to have any excess flux scraped off, and checked for any open circuit heaters. This is to be done by monitoring the current with a clamp on ammeter after removing the cover. The current in each heater block is to be 5amps.
The solder pot heaters are to be checked in the same way. The current for each heater is also 5 amps.
The conveyor is to be cleaned, and any flux scraped from the fingers. The speed is to be checked over the marked length, with the controller set on the calibration mark. It shall not deviate from 6fpm by more than 1%.
The conveyor is to be greased with Hi Max grease by applying with a brush to the entire length of the conveyor chain.
After completing the maintainence Q.C. shall make a check of all operating parameters.

Monthly Maintenance. Once each month the machine shall have the covers removed from the conveyor, and the fluxer removed. All mechanical parts shall be cleaned, checked for wear, lubricated, and adjusted. All heaters and motors shall be checked for insulation resistance and power consumption. The fluxer shall be cleaned and checked for performance. Any worn parts in the system shall be replaced. On completion Q.C. shall check the system to the performance specification.

Reporting. All maintenance shall be reported to the S.P.C. with details of any defects found, or any parts replaced or adjustments made.

Solder Dumping. When it is required to dump the solder it shall be drained into molds, the pot cleaned of any accumulations of oxide or dross, and refilled with melted solder from the gas heated pot stored adjacent to the solder machine. Do not use any tools on the pot except those provided for the purpose.

Figure 6.8. A simulated document describing exactly the maintenance requirements for the solder machine.

out and recorded correctly. In addition, the instrumentation of the solder machine itself must be checked on a regular basis, including any measuring devices used in the process, for example, the hydrometer used to manually monitor the flux density. This checking must be carried out at least once each week, and even if it proves that the machine is functioning correctly, must not be omitted from the regular system of process control. Most companies already have this form of monitoring as part of the QC organization, and there is usually no need to do more than ensure that the machine instrumentation is included in the procedure. For the smaller company, where no such system exists, it is not difficult to repeat the simple checks that were described in Chapter 5 in the section Putting It All Together. A good thermometer and a watch with a second readout will be adequate for the simpler machine, and the few minutes spent to make this check once a week and to record the results will be well worthwhile if only to avoid greater problems.

It may appear that these comprehensive checks and detailed instructions and record keeping are an unnecessary burden. In the following chapters it will be shown that they are vitally important, and that it is too late to put them into operation when a problem arises.

Chapter 7

Zero Defect Soldering

Zero defect soldering is not a new idea. It was first discussed several years ago, and has been revived many times. Today, with the large densely packaged boards, and the emphasis on quality and reliability, it is slowly but surely being accepted and initiated by the more progressive elements in the electronics industry. This philosophy states simply that the only way to arrive at a completely reliable soldered printed wiring assembly is to control the mass soldering process so tightly that 100% excellence in soldering is ensured, and no touch-up is ever necessary.

Zero defect soldering is also much lower in cost than any other method of assuring a faultless assembly, in spite of the additional control of the materials and processes required.

As discussed in earlier chapters, mass soldering replaced the hand soldering operation and produced an enormous improvement in productivity. The fact that the machines and materials were not perfect did not cause any major problem, for with such a spectacular reduction in labor expenditure, the use of a few people to inspect and rework any improperly made joints was only a minor irritation. Small boards were also the order of the day, with comparatively simple circuitry, and inspection and touch-up were quite feasible.

Machines improved, better solders and fluxes were developed, but the solder joint inspector and the touch-up operator continued as an automatic and inevitable part of the operation. Not to be used as a backup for deficiencies in the materials or machines, but to cover a lack of process control in the operation, inept accounting when the cost of the soldering process is calculated, and frequently a lack of understanding by management of the nature of the soldering process. These factors and the claims of the zero defect soldering philosophy will be examined in detail.

THE COST OF SOLDERING

More than any other factor, this one aspect of soldering is generally totally misunderstood, especially by those on the periphery of the process. Some of the arguments that are frequently heard run as follows:

> If we buy lower-grade solder we can save three cents a pound, and that will amount to $5000 for the year.
>
> The flux only needs to be dumped every third week, which will save $5000 a year.
>
> If we put a solderability specification on the purchase order the component price will go up 2%.
>
> It will cost $2000 to change the design to get rid of the problem.

The basic assumption here is that soldering is a cheap process, and if there is a problem it only means putting on another touch-up operator. This, of course, is a total fallacy brought about by the incorrect calculation of the true cost of faulty soldering. Soldering is a low-cost process, but faulty soldering is one of the most costly processes in the electronics industry. When the cost of making soldered joints is considered, the evaluation must stretch further than the actual soldering operation itself. The cost of ensuring solderable parts, the cost of any touch-up and inspection, and, of course, the ultimate expenditure of putting right any field failures must be included in the final figure.

When all these factors are put together and the calculation correctly done, several startling facts become apparent. If the $5000 saving from buying a lower-grade solder results in only a very tiny increase in solder joint failure then the savings will be nullified, and indeed the overall cost of the process will increase.

FAULTY SOLDERED JOINTS ARE VERY, VERY, EXPENSIVE. Almost anything that will reduce the number of faulty joints will show a cost saving. Remember that the concern is not with bad soldering, but with soldering that is generally considered quite satisfactory. An example is interesting, perhaps a little frightening. This costing is not of any particular company, but was put together with data obtained from several sources. It is, however, realistic and typical of many operations.

A plant is making 10,000 units per week.	
Each unit has 1500 soldered joints.	
Solder joint failure as measured at the solder joint inspection is 0.05%	
Total number of faulty joints per week	7500.
Of these 75% are found and repaired at touch up at $0.50 each	$2812.50

Of the 1875 faulty joints left, 50% are repaired at final test at $5 each	$4690.00
Of the 937 faulty joints left, 75% cause field failures and are repaired at a cost of $85 each	$59,670.00
Total cost of the 0.05% solder joint failure rate	$67,172.50 p/w
If each field failure occurs in a separate unit, the field failure rate will be	7%.

When it is considered that a solder joint failure rate of 0.05% is frequently considered good, the overall cost of bad soldering in the electronics industry is nothing short of phenomenal. Of course, the problem is that the normal accounting procedures usually put field failures into a separate budget from that of the factory, and so the true costs are never seen. It may be very instructive to put your own data into a similar calculation. In fact, this is often necessary in order to convince management, design engineering, purchasing, and other authorities that the steps demonstrated in this chapter are not only necessary but are extremely cost effective when the overall figures are considered.

In the example shown a solder joint rejection rate of 0.05% costs well over three million dollars per year to rectify the defects. Even worse it results in a field failure rate that can only damage the reputation of the product, and eventually the profitability of the company. This clearly demonstrates the futility of setting a so-called "acceptable failure rate." The objective in soldering as in any other process must be to eliminate any failures. Zero defect soldering is the only cost effective philosophy that can be accepted in the industry.

SOLDER JOINT RELIABILITY AND TOUCH-UP

There have been many suggestions that touch-up does not produce joints that are as reliable as those made during the first soldering operation. There is also considerable supporting evidence in the form of tests carried out on both production and test boards by thermal cycling, vibration, and other forms of environmental stress. As with all soldering tests they are quite difficult to carry out because of the many variables involved, and the large lot sizes required to ensure that the final data are meaningful. However, all the evidence together appears to be much more than circumstantial, and there is general agreement that the uncontrolled use of the soldering iron in the hands of the rework operator can result in a joint with a shorter life than one made in the controlled process of the soldering machine.

The fact that the joint did not solder satisfactorily in the machine will probably require the operator to apply more heat, or make more than one application of the soldering iron, in order to overcome the lack of solderability that prevented the formation of a sound joint in the first place. In other words the faulty joint can fail for two reasons: first, the problem that caused the joint failure during the machine soldering, be it poor solderability of the parts, bad joint design, high thermal mass, or whatever; and second, the handwork required to solve the initial problem.

The answer surely is to ensure that the soldering process and the condition of the parts are such that the joints are correctly made at the initial machine soldering.

The concept of zero defect soldering, therefore, offers the ultimate in solder joint reliability and quality, at the lowest cost. This is not to say that zero defect soldering requires the acceptance of anything other than the highest level of solder joint specification or physical appearance. There have been efforts to make solder touch-up unnecessary by reducing the requirements for smooth fillets, complete filling of holes, and other indications of excellence in soldering. This is not the philosophy of the zero defect soldering process. Indeed, the concept says that any variation from the perfect joint indicates that there is something wrong with the machine, materials, or process parameters, and that the only way to produce good soldered joints is to maintain total control of these factors.

SOLDER JOINT INSPECTION—THE IMPOSSIBLE TASK

The touch-up of soldered joints requires that the faulty joint is identified. This is either left to the touch-up operator, or to an inspector who marks the joints that have to be reworked. This is an extremely subjective process, and everyone who has been involved in this work will have experiences that demonstrate just how subjective this inspection can be. The author was involved in such an incident some years ago. The number of touch-up operators on a particular line was steadily increasing. They inspected and touched up the assemblies, which were then sent for final QC inspection. Invariably a considerable number came back for some joints to be reworked a second time, or for joints to be touched up that the operator had considered satisfactory. As a result, the touch-up operators reached the point where they were reworking almost all the joints to avoid having the assemblies returned twice, and sometimes three times.

The operation was changed so that the inspectors marked those joints that had to be reworked, and the assemblies were then sent to the touch-up operators for this work to be carried out. After rework the assemblies were cleaned and sent to inspection for a final check. Once again, some assemblies were returned for further rework, and as the weeks went by the amount of touch-up once again reached unacceptable proportions.

At this point a test was set up. One of the assemblies that had been inspected was removed from the line and photographed. The markings were washed off, and the assembly introduced into a following lot. It was inspected, and again photographed. After touch-up it was cleaned and once again introduced into another lot for inspection. After this third inspection it was once again photographed. When the three photographs were checked side by side there was no doubt as to the randomness of the inspection results. Some joints that were considered satisfactory on the first inspection were required to be reworked at a subsequent inspection; some that were reworked were found once again to be faulty; and so on. The subjectivity of the entire process was very evident. The cure was to train the inspectors very thoroughly, and this training had to be repeated every month or the situation fell back into its old ways. This is not to blame the inspectors; they were all trying very hard to make objective judgments in a situation where the very inspection parameters are written in subjective terms.

If the actual inspection process is considered, the inspectors minds are littered with such things as

Too much solder
Insufficient solder
Frosty joint
Cold joint
Disturbed joint

Not to mention the pin holes, voids, and other real and imaginary defects. Not only is the inspector required to decide if a joint requires rework, he is expected to define the nature of the failure and the reason for rejection. Perhaps our products would be of higher quality if our inspectors were shown a perfect joint and told that "Only this is acceptable. Everything else must be considered faulty and indicates that there is a failure in the process, which must be stopped until the problem is found and corrected." There is, after all, little sense in continuing to produce a faulty product, which will require to be reworked, and as a result will only suffer further degradation.

The boast is often made that 100% inspection is carried out on all soldered boards, inferring that every joint is inspected. This, of course, is rarely the case. The cost of this level of inspection would be prohibitive, and the chance of missing some joints would be very high unless some kind of inspection fixture or machine was used. Even suppose that such a device was available, it is only the exterior of the joint that is visible, and inspection does not guarantee that the internal structure is satisfactory.

For all normal levels of solder joint inspection the inspector is at best sampling the quality of the joints on each board, and from the external appearance of the joints is assessing the chances that a satisfactory bond has been achieved on all of them. This assessment of each and every joint is obviously ridiculous. The responsibility for perfect joints lies with the process, not with the inspector. He is only valuable in providing the first warning that the process is getting out of control.

In order to make the point of course, the above paragraphs have somewhat simplified the situation. It is necessary to use some touch-up from time to time. When a board is found with a joint that has not soldered, the product cannot be shipped in that manner, and a hot iron and some cored solder is probably the best way to make the repair. But the first action must be to find out why it happened, and to prevent a repetition of the fault. The touch-up operator must not be considered a part of the production team, but should only be used on the rare occasions when the process has failed, and the products that have been affected by the failure have to be made satisfactory for shipment.

Similarly, the solder joint inspector is necessary and is required to identify the various forms of solder joint failure. Not to be able to log them onto a failure report, but to be able to inform the process engineer, and so trigger the first actions to solve the problem. It must be recognized that the inspector cannot and must not be held responsible for every joint that passes by him, but that he is the early warning mechanism to monitor the performance of the process.

The types of solder joint failure and their probable cause will be discussed more fully in the section on Troubleshooting the Process. The inspector will need to be able to identify the various classes of failure, and regular training is necessary if this is to be carried out effectively. A good set of slides or photographs will be found to be extremely useful for this, and these are available from several sources. The Tin Research Institute has an excellent set of slides, and the documents available from the IPC are also good training material. One of the best methods of maintaining a constant level of inspection, and consistent terms for the various defects, is to prepare an in-house exhibit of typical failed joints. This must have full information on each defect, including a description of the visual criteria that is used to define the failure, and the suggested troubleshooting routine. While slides and photographs are useful tools, the three-dimensional examples are much easier to use. The words used to describe the failure should be agreed on by all concerned so that a common set of terms is used throughout the organization. They should help to identify the cause of the defect and the corrective actions that must be taken to eliminate the problem. For example, the term "insufficient solder" is meaningless while "blow hole" or "nonwetting" will immediately indicate the appropriate area of the process where the cause will be found. To sum up the area of solder joint inspection:

It is impossible to ensure sound solder joints by inspection.

Solder joint inspection should only be used as the first warning of an out of control process.

Inspection cannot be used as a replacement for process control.

Inspectors require effective and continuous training.

Every effort must be made to establish and maintain agreed inspection standards.

At the first sign of an increase in faulty joints, the inspector must shut down the process until the cure is found.

Touch-up must only be used as a temporary measure to repair faulty work, never as part of production.

Any joint that is not perfect is a sign that the process is getting out of control.

SOLDER JOINT INSPECTION MACHINES

With all of the difficulties associated with visual inspection, it was to be expected that attempts would be made to automate the process. Several machines have been developed and marketed. They generally fall into three main classes.

1. Thermal systems in which the joint is heated rapidly by a laser pulse and the amount and rate of heat loss is then measured. This produces a "thermal profile" of the joint from which the quality of the solder bond can be deduced by comparison with the same data from a known good master board. This system has been developed into a very practical machine and is probably the most widely used and tested type (Figs. 7.1 and 7.2).
2. X-ray systems which produce the typical interior view of the joint either on a photographic film or on a cathode ray tube monitor, or on both. The image has to be reviewed either visually or by computer analysis. This inspection method has been primarily used to inspect surface mounted joints where the fillet is not easily visible and can clearly identify voids and lack of solder.

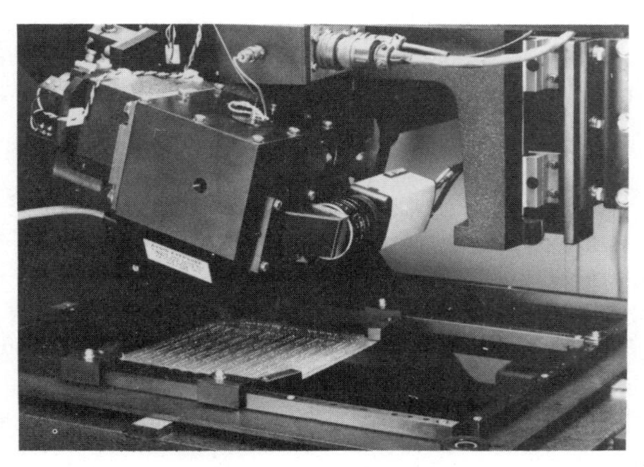

Figure 7.1. The optical head of a laser I.R. solder joint inspection machine. Courtesy Vanzetti Systems.

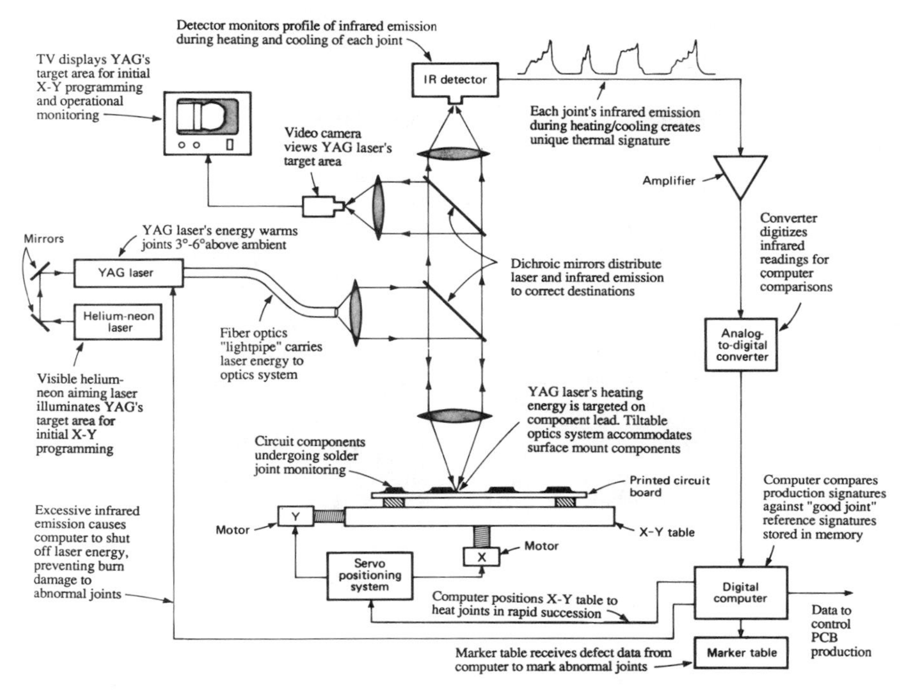

Figure 7.2. A schematic diagram of the functioning of a laser I.R. solder joint inspection machine. Courtesy Vanzetti Systems.

3. Computer controlled vision machines in which a T.V. camera produces a magnified image of the exterior of the joint. This image is then digitized and the computer compares the digitized signal with data held in memory. This form of inspection of course is basically the same as that of the human inspector except that it removes subjective opinions with respect to the joint quality and does not suffer from tiredness or boredom (Fig. 7.3).

All of these systems have their problems. Inevitably a very small number of defective joints will be defined as satisfactory and a few good joints will be identified as defective. The exact number and nature of these will depend on the type of system used, the type of joint and the form of the assembly packaging. These machines do not get away from all of the problems of visual inspection, but they are consistent.

As previously discussed solder joint inspection should not be used to sort out good joints from bad, in an attempt to correct defects already built into the product. The emphasis must be on making the product right the first time. Because of their tireless and consistent operation these machines are ideal for providing the data necessary to control the soldering process. In

Figure 7.3. A typical computer controlled vision system for inspecting solder joints. Courtesy Benchmark Industries Inc.

this function there is no doubt that the machines will eventually replace the human inspector. In addition most will analyze the results of the inspection and place them into a suitable format for use in the control of the soldering process.

Microsectioning Joints for Inspection

When solder joint problems occur it is often not possible to determine the exact failure mechanism by inspecting the exterior of the joint. Indeed, it is often not known if the joint is faulty or not. For example, although dull and frosty joints are not desirable and immediate actions must be taken to rectify the problem, from their external appearance it is impossible to determine if the joints already made will jeopardize the reliability of the product.

Or again, when dewetting occurs of the PWB or the component lead, it is not a simple matter to decide precisely on the reason for the problem from a simple visual inspection.

A microscopic examination of the joint is the first course of action, and frequently this alone will give a clue as to the problem, and thus the cure. However, ultimately it becomes necessary to look further and examine its internal structure. This is not difficult to do, and is carried out by making a cross section of the joint and inspecting it under the microscope. With suitable magnification, of up to 200 times, all of the internal structure of the bond can be seen, including the intermetallic compound which will confirm wetting, the crystalline structure of the solder, and the other metals of the joint.

By now, the usefulness of this method of inspection will be apparent from the many examples in this book. The necessary skills for using this technique are easily learned, and the cost of the equipment to make and examine sections is not excessive, and will repay its cost many times in the ability to assess accurately the actual nature of the solder joint.

In practice the part of the PWB containing the joint to be inspected is cut out and embedded in a plastic compound in a suitable container. A thermosetting material is sometimes used to encapsulate the sample, or an epoxy for the so-called cold setting mounting of the part to be sectioned. When the compound is hard, it is removed from the container, and is now usually in the form of a short cylinder, about 1 in. (2.5 cm) in height and diameter. This cylinder is then ground away to expose that area of the joint that it is desired to examine. The hardened potting material supports the joint structure during the grinding and subsequent polishing, so that the very fragile parts are not distorted or damaged in any way (Fig. 7.4).

A good microscope designed to be used for inspecting these metallurgical specimens is the most costly part of the necessary equipment. However, this and the grinding materials are the only things required to carry out effective microsectioning. The grinding can be done manually if only an occasional section is to be made, or motor driven grinding wheels can be obtained at a very reasonable cost. The necessary skills can be self taught from one of the manuals on this subject, but a few hours of instruction will help in attaining the skills more quickly, and also introduce the newcomer to some of the "tricks of the trade." Classes can often be found at the nearest technical college, and those run by the IPC about twice each year are well recommended. Almost any printed wiring board fabricator uses this technique, and will often give guidance and advice.

Under the microscope it is also possible to measure the parts of the joint.

Figure 7.4. Some typical encapsulated joint samples ready for cross sectioning and a ring mold used to pot the samples.

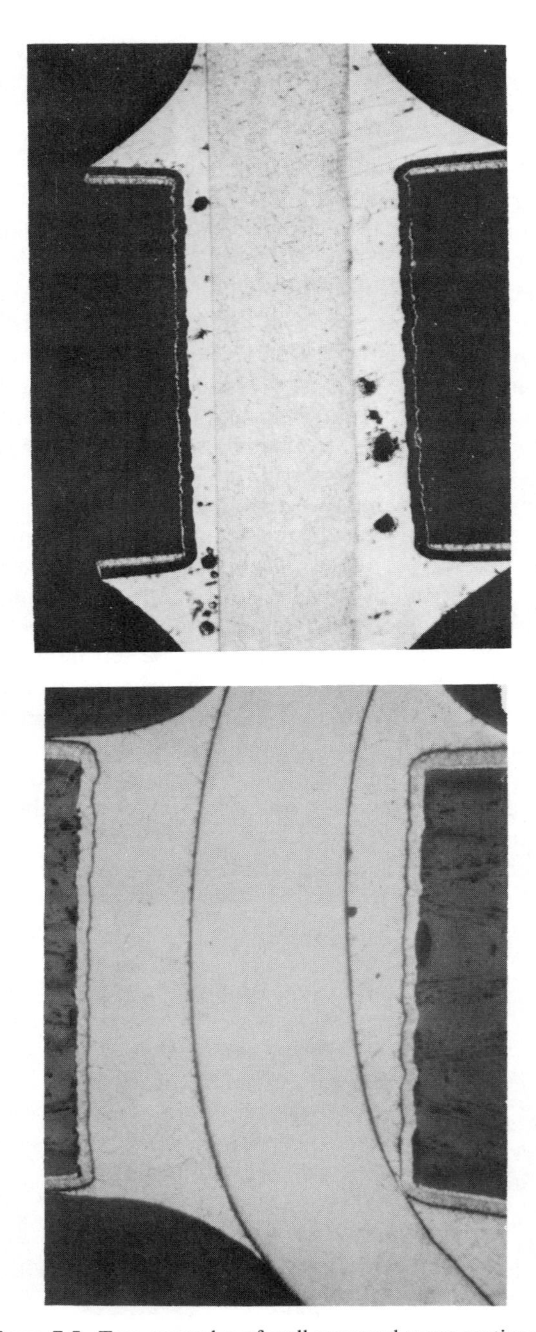

Figure 7.5. Two examples of well-prepared cross sections.

The intermetallic thickness and that of the copper foil or plating can be checked. When the solder does not form a good fillet on the top surface of the board, a microsection will quickly show the thickness of the through hole plating and any cracking that may be causing the problem. Voids and pin holes can be inspected to see if they are indicative of more serious problems or are merely unwanted cosmetic details (Fig. 7.5).

Most microscopes have attachments for a camera, so that a permanent record can be made for future reference. Microsectioning is the most useful method of solder joint analysis that exists, and anyone who has the responsibility of running a soldering process should have available the equipment and skills to be able to make full use of the technique. Indeed microsectioning is the only way to really tell if a joint is sound or not. Unfortunately, it requires the destruction of the joint and so the emphasis must be placed on controlling the process and thus making the inspection of the joint unnecessary except for use as an investigative tool.

SOLDERABILITY—THE KEY TO SUCCESS—THE 70% PROBLEM

In almost every plant, if the solder machine is correctly set up and operated and the causes of defective joints are analyzed the following approximate ratios will apply.

Solderability of boards	50
Solderability of components	20
Handling and assembly	15
Design defects	5
Solder machine	5
Others	5

It can be seen therefore that excellent solderability of the parts to be joined is the single most important factor to be considered. It also explains why attempts to solve all soldering problems at the solder machine are doomed to failure and eventually generate the idea that soldering is some kind of magic or art. In fact of course it is a totally controllable and repeatable process.

Solderability has been discussed, reported on, and written about for many years. There is really no reason why every soldering operation should not start with good solderable parts. Most often the problem is not found to be technical at all but one of ill-informed economics, poor organization, or a lack of communication within the company. As an example, there is a reluctance to change from bare copper boards to solder coated boards in cheaper products, on the grounds of cost. Yet if an analysis is carried out as in the previous chapter, it will almost certainly be found that what appears to be a saving is, in fact, causing additional expenditure.

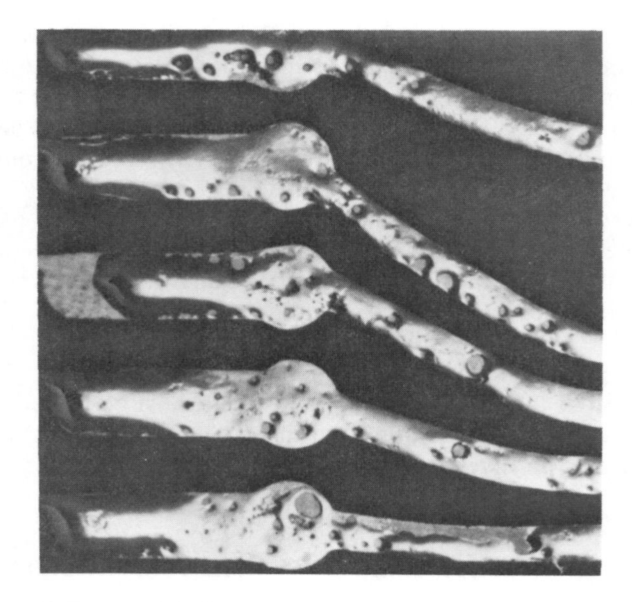

Figure 7.6. Two examples of defective solder joints caused by poor solderability of the printed wiring board. Courtesy Tin Research Institute.

Similarly, solderable components that have been hot tinned can be obtained, but may require additional effort on the part of purchasing, or a slightly higher cost. First in, first out storage is definitely required, but here again this is sometimes not acceptable to the organization on the grounds that it is more difficult, or requires more people, and so on.

The question of solderability begins, therefore, not with hardware, but with the realization that the mass soldering process is extremely low cost if it is correctly controlled, but can generate an enormous expenditure if it is not. Solderability is the key to making consistently good soldered joints, and eliminating the cost of identifying and repairing faulty ones.

Solderability is the ability of the metallic surface to be wetted with solder. As discussed in earlier chapters, this is a matter of cleanliness and the action of the flux. The more active the flux, the easier it is to wet the surface. It is sometimes suggested that the more active fluxes do not require the same solderability to make a good joint. Although there is certainly some truth in this idea, it is dangerous to rely on the flux to take care of poor solderability. No matter which flux is being used, the mass soldering of electronic assemblies requires the very best solderability of the parts if it is to be successful (Fig. 7.6).

Solderability Testing

All parts to be used in the mass soldering process must be tested for solderability before being accepted from the vendor. The purchasing specification must contain a solderability requirement, and QC must monitor the parts to be sure that this requirement is met. One very simple method of organizing this is as follows:

- Each part number is placed in a file, one file per part number per vendor. If a computer is available it will provide a quick and simple system of maintaining these records.
- When a part is received a sample is tested for solderability. Any normal sampling plan can be used but common sense is probably the best plan of all.
- The results are logged into the appropriate file.
- Feedback from the process control solder joint inspection station, with respect to solderability problems found on the shop floor, is added to each file.
- From the results of solderability testing and shop floor inspection, a history of each part number per vendor is generated. From this history the sampling rate at solderability testing can be increased, reduced, or eliminated.
- Once accepted, parts must be used on a first in first out (FIFO) basis.
- Parts that have been stored for six months must have their solderability tested again and each following six months of storage.

This system reduces the solderability testing to the absolute minimum, yet retains control on all the incoming parts. Sometimes some of the parts are produced in-house, especially PWBs, and there is a tendency not to test these because it is felt that being in the same company this will be a duplication of effort. This is wrong; it is difficult enough to make the people involved realize the importance of good solderability, and it is not likely that the PWB shop will see this requirement as anything more than of secondary importance. The only exception to this rule is when the boards are hot air leveled, as this process is virtually a solderability test in itself and any deficiency is immediately visually identifiable. The secondary aim of all this testing must be to communicate any unsatisfactory results to the vendor, and assist him to improve his product to the point where testing is nothing more than an infrequent monitoring of his performance.

There are many different ways of carrying out solderability testing, and testers are available for this purpose. These fall into three main categories:

Those that measure the wetting time of a wire by timing the rate that a globule of solder reforms after being split by the wire being tested (Fig. 7.7)

Those that use capillary pull principle, and measure the solderability of the specimen from the buoyancy and wetting forces acting on it (Fig. 7.8)

Those that dip the part into solder and then rely on a visual inspection to determine the solderability; the dipping can be manual or automatic (Fig. 7.9)

All these systems are perfectly satisfactory, but the first two give a numerical value for solderability, and take away the subjectivity of the visual test.

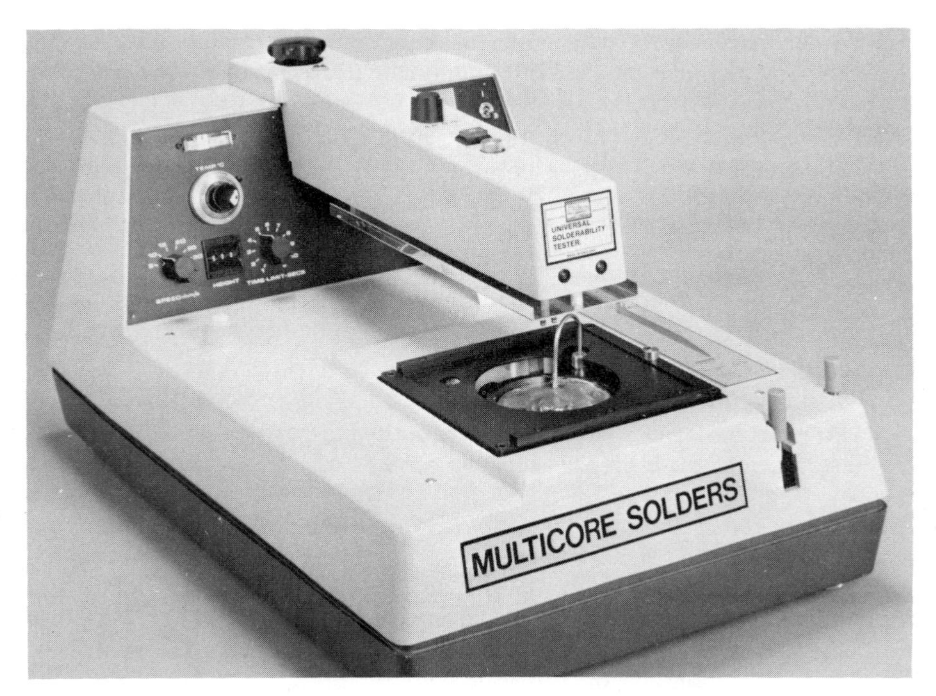

Figure 7.7. A universal solderability tester which will carry out the globule, wetting balance, and the simple dip test. Courtesy Multicore Solders Ltd.

Figure 7.8. An example of the wetting balance solderability testing machine. Courtesy Hollis Engineering.

However, the visual tests are fast, and can be carried out with nothing more than a solder pot and a jar of flux. With a skilled inspector they will be extremely accurate.

The globule tester is in theory a very simple device, although it has been developed into quite a complex machine. A solder pellet of a predetermined weight is melted on an electrically heated block to form an almost spherical globule. If the wire to be tested is now fluxed and placed on the globule it will instantaneously split the solder into two segments. If the wire wets, the solder will flow around it, and reunite over the top (Fig. 7.10). The time taken to do this is a measure of the solderability of the wire. The timing can

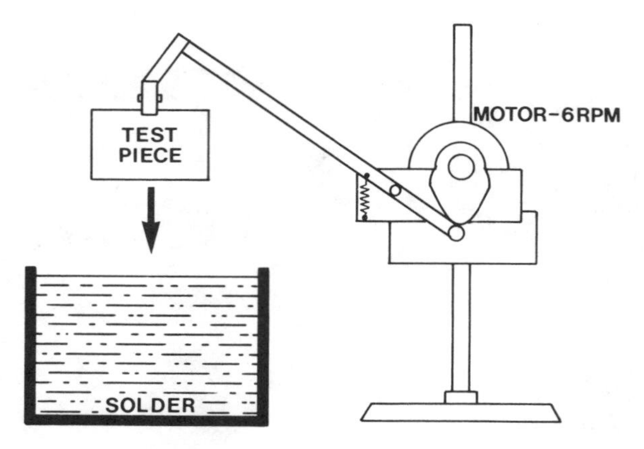

Figure 7.9. The edge dip tester.

be done visually by watching the wetting of the wire through a microscope, or automatically by using a contact point mounted just over the top of the wire, which is activated by the molten solder as it completes encircling the test sample.

In some testers the same method can be used to measure the rise time of solder through a plated through hole. The probe in this case is placed over the top component mounting pad, and is activated by the solder flowing out over the pad surface. The globule test is described in detail in the following international and European specifications: IEC 68–2-20, BS 201 1, and DIN 4006.

As these specifications suggest, this test method, although a perfectly satisfactory way of measuring solderability, is much more commonly used in

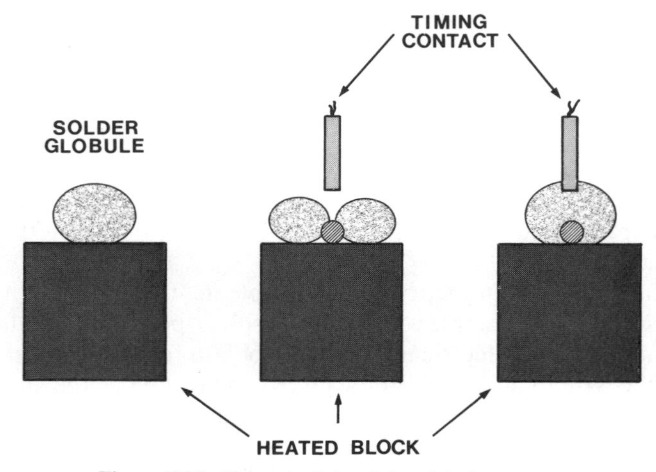

Figure 7.10. The principle of the globule tester.

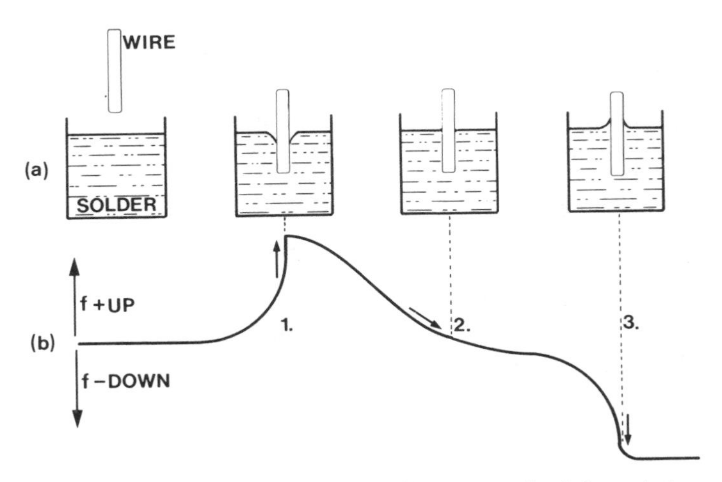

Figure 7.11. The principle of the capillary or wetting balance test.

Europe than in North America. It has certain limitations in that it can normally only be used on circular section leads, and the size of globule used has to be quite precise and varies with the size of the wire to be tested. The chief objection is that the portion of lead tested by this method is very small, and may not be representative of the solderability of the entire lead. The advantages of the globule test include the fact that to some extent this simulates the actual machine soldering process, and most important that it does produce a definite numerical measure of solderability. Although normally only circular leads are tested on the globule tester, by careful control of the fixturing on the machine, other shaped leads, for example, from DIPs, have been successfully tested for solderability.

The capillary pull or wetting balance is a very elegant method of measuring solderability. As shown in Fig. 7.11, when a sample is immersed in molten solder, it will displace some of the metal, and there will be applied to it an upward or buoyancy force. If the sample is solderable, as it heats, the solder will wet and form a convex fillet around it. Now the forces are reversed. The capillary force forming the solder fillet will tend to pull the sample downward.

By mounting the sample on a suitable balance, the direction and magnitude of the forces can be measured and plotted with time. This gives a complete picture of the solderability of the part. In the modern versions of this form of tester, the parameters are monitored by a microprocessor, which computes the percentage of complete wetting and also the time to reach 95% of total wetting. This method of solderability testing also gives a numerical value for the results, and does test a larger portion of the component lead.

The tester is quite complex and comparatively expensive. It is, however, capable of analyzing the actual wetting of the part in great detail, and provides the maximum amount of information on the way that the part will actually perform when it passes through the solder wave. The manual dipping test methods are the simplest to carry out, and are generally accepted by many specifications: for wires, Mil Std 202, Method 208; for PWBs, Mil P 5510, IPC-S-804.

The tests can be carried out manually, although for some, the so-called *Edge Dip Solderability Test* of IPC 804, for example, an edge dip solderability tester is required. This is a very simple piece of equipment, which can even be made quite easily in any workshop with hand tools (Fig. 7.9).

For manual testing the parts are dipped into water white rosin flux, Type R, and then dipped into solder at a defined temperature for a specific time. The leads are then cleaned and inspected. For acceptable solderability the leads must have a defined coverage of smooth, shiny solder. For Mil Std 202, for example, the coverage must not be less than 95%, and any defects must not be collected in any one area. It is, however, illogical to demand excellence in soldering and less than excellence in solderability. There is therefore a move towards requiring 100% wetting of all component leads. For some tests, the leads must be artificially aged before testing. The product specifications should be referenced to see if any specific solderability tests are required.

For PWBs the edge dip tester is called for. The PWB has a section cut from it, which is mounted on the arm of the tester. The arm dips the board sample into and out of the solder in a predetermined path and at the specified speed. The board is then compared with the photographs given in the IPC manuals, IPCA-600c, or IPC-S-804. Where no particular specification is called for by the product requirements, those mentioned above provide excellent starting points for initiating a solderability testing program.

In such a program there are several factors to be considered. As previously mentioned, the aim must always be to arrive at such an excellent understanding with the component vendors that solderability testing becomes a monitoring function only. Any tests, therefore, must be such that they can be used both in-house, and by the suppliers, so that both parties are using a common solderability measurement. The number and variety of parts must be considered, is well as the cost of testing. Where justified, both the globule tester and the wetting balance will perform well, and the results between different machines can be correlated easily because of the numerical measure of solderability from these two systems.

The dip test is more subjective, but, of course, is lower in capital cost, and somewhat faster to carry out. The disadvantage of the method is the subjective interpretation of results, especially where the part is marginally near the specification limits.

Even with small numbers of parts there is no excuse for not performing solderability testing. Ajar of R type flux, and a solder pot about 6 in. (15 cm) in diameter, with a suitable temperature control is all that is required.

Fill the solder pot with 60/40 solder and operate at 470°F (243°C).

Maintain the flux at the correct density, and keep covered when not in use.

Dip the sample to be tested into the flux.

Allow the surplus flux to drain off.

Scrape any dross off the solder surface, and then immerse the part for 4 seconds; withdraw and allow it to cool.

Wash with alcohol, and examine with a 10 power magnifier.

The lead should have a smooth, shiny coating of solder without pits, spots or any other defect.

Any voids or nonsoldered spots that are seen should not exceed 5% of the total surface soldered. It is extremely difficult to assess 5%, and anything that can be seen with the naked eye should be the cause for testing more samples and for rejecting the lot if seen other than extremely randomly. The examples and photographs in Mil Std 202 (Figs. 208–6 and 208–7) will be found to be extremely useful.

This is a very simple test, and with a little experience and training can be carried out quickly by almost any operator. The same test can be used to check the solderability of the surface of PWBs by cutting a strip about 1 in. wide from the board, and dipping in the same way as the lead on a component. To avoid solder splashes the flux should be dried for a few seconds by holding the board over the hot surface of the solder prior to dipping. This test will check the surface of the board for solderability, but if the board has plated through holes these also must be tested.

The so-called capillary or wicking test will provide complete information on the way that the holes will behave during soldering. To carry out this test, cut a small square from the board, about 2 in.^{2} (5 cm^2), containing as many holes as possible. Dip in the flux, allow the surplus to drain, and dry for a few moments over the solder surface. Skim the dross from the pot and float the board on the solder. Time the rise of the solder in the holes and up over the surface of the top pad. If it takes more than 5 seconds the board is suspect; more than 10 seconds and it should be rejected. Experience shows that this test is not time dependent.

This test not only checks the actual solderability of the metal in the holes but also the integrity of the through hole plating, and indicates any major plating voids or cracking which will produce voids in the solder filled holes.

Both of these manual tests are simple, fast, and with some little skill and expertise are extremely effective. However, they must be used with understanding; for example, the latter test will require the times to be changed when multilayer boards are checked because of their greater thermal mass. Similarly, larger holes may not permit the solder to wick up as fast as smaller holes.

This may sound rather complex and somewhat ill defined for a test specification. In fact, once some testing has been carried out it will be found to be extremely simple to set test parameters. Exact figures for some tests cannot be given because of the many variables involved. The objective is to make tests that will allow components to be rejected that will not solder readily at machine soldering, and the act of soldering can always be used as the final referee.

If there is any doubt as to the validity of any of the testing, take some of the components from both passed and failed lots and actually run them over the solder machine. Compare the results and then make any adjustments necessary to the test requirements.

Wherever possible, relate the test to one or other of the national solderability specifications applicable to the product being assembled; this saves a lot of work and ensures that the vendors will accept the test results. It costs comparatively little to set up the equipment and organization necessary to ensure solderable parts. The cost will be saved many times over in the reduction of solder joint inspection and touch-up.

Solderability Problems

Printed Wiring Boards (PWBs) are a particular source of solderability problems. As previously mentioned, the hot air leveled board is really the total solution to the problem, because in that process the board is dipped into flux, solder dipped, and then subjected to a blast of hot, compressed air. Any areas that will not wet are immediately visible, and any defects are very obvious. With this finish a visual inspection is all that is necessary, and once processed the boards have an extremely long shelf life, and require only normal care in storage and handling. No other finish can compare with these qualities.

However, the hot air leveling process must be correctly carried out and controlled, with very careful cleaning of the basis copper and correctly set up machine parameters. If too much solder is removed by the air knives the exposed intermetallic compound at the knee of the hole can make it difficult or impossible to achieve a top fillet when a plated through hole (PTH) board is soldered. This defect is known as "Weak Knees" (Fig. 7.12).

Any of the plated finishes require testing for solderability, and then careful handling and storage; some have a limited storage life and will require retesting before soldering.

Boards that have been solder plated and reflowed offer the next best solderable surface, and if processed correctly are as good as those that have been hot air leveled. The catch is in the phrase "correctly processed." If the solder has been plated over copper that has poor solderability because of oxide, dirt, or incorrect plating, it will still reflow, and because it is not subjected to the forces of the air knife, can produce a smooth shiny surface that has all the appearances of excellent wetting. Boards with plated and

Figure 7.12. A typical example of a "Weak Knee." The solder coating at the knee of the hole is so thin that it has all been transformed into intermetallic compound and is not solderable.

reflowed solder must be tested for solderability. When they have been passed as satisfactory they will need no more special treatment than the hot air leveled board.

One other method of applying a solder finish is by roller tinning. In this process the board is passed between two rollers, one of which dips into a bath of molten solder. The solder is transferred from the roller to the board surface. This is an extremely old system and was very popular in the early days of the single-sided board. However, as with the plated board the success of this process depends on the solderabiity of the copper surface onto which the solder is deposited. It is possible to produce excellent looking coatings, smooth and shiny, on surfaces that will prove to be of very poor solderability once the solder coating has been melted in the solder machine.

All roller tinned boards must be tested for solderability. This form of solder coating also produces a layer of solder that is thin and variable in thickness, and a clearly limited shelf life must be imposed on these boards which will be dependent on the storage conditions.

Plated finishes require excellent process control when they are applied if the result is to be satisfactory. All must be suspect until they have been proved solderable by testing. In addition they are frequently thin, and have a short shelf life. There is no way of determining visually the state of the surface to which the ultimate solder bond has to be made. Plated finishes are perfectly satisfactory when correctly processed, and this is one area where there must be a good understanding by the vendor of the importance of plating the specified thickness of metal over clean solderable copper. An adequate test procedure should be agreed on with the vendor, and he must be required to carry out these tests and send the results with each batch of boards shipped. It is expensive for the board vendor to make boards and then have them rejected. It is unprofitable and inefficient to receive and test boards only to find that they do not have excellent solderability. This is another case where close cooperation between vendor and customer can pay dividends for both parties.

Finally, let us consider the bare copper board. If plating and cleaning have been correctly carried out, the board will have excellent solderability. The shelf life will be short, and the board must be protected from the effect of exposure to the air, and any handling. These boards should always be solderability tested immediately before using.

Copper boards are sometimes coated with a lacquer in an attempt to retain solderability. This coating is usually rosin in alcohol, that is, a water white flux. As discussed in Chapter 3, rosin is an excellent flux for the removal of copper oxides. Unfortunately, it is not active until it is heated and melted, and after some time on the board surface during storage it becomes extremely hard.

Flux is applied to the board before soldering, to clean off the contaminants and oxides from the board surface and the component leads. With the lacquered board the solder will wash off most of the flux from the board, which will have been protected from its action by the lacquer. For the board, therefore, the only flux will be that provided by the lacquer itself. As has been noted this is not a very active flux, and being hard from storage will not melt readily and become as chemically active as freshly applied flux.

For this reason, it is often found necessary to wash off the lacquer with a suitable solvent such as alcohol to allow a more active flux to be used. This finish, therefore, should always be suspect, always be tested for solderability, and always retested immediately before use.

There is therefore a wide range of PWB finishes, and this is definitely an area where the lowest-cost finish can be the most expensive in the long run.

In the same way that board finishes affect solderability, the terminal and lead finish on the components play a major part in achieving zero defect soldering. The solder coated lead produced by dipping the lead into flux and molten solder (hot tinning, as it is called) is the most solderable, and has an almost indefinite shelf life. Plating of any kind, as with the PWB, must be suspect, and must always be tested for solderability. Semiconductors with iron based leads are particularly troublesome, as once the solder removes the plating, none of the normal fluxes will allow wetting of the base metal. Fortunately, many manufacturers are turning from plating to hot tinning, which removes the problem. It is not that plating of itself is an unsatisfactory process, but without strict control it is possible to plate onto a base metal that is contaminated, and produce a finish which appears perfectly satisfactory. When the lead is soldered the plating is removed by melting or dissolution in the solder, and the base metal will be totally unsolderable. When the base metal is one of the iron based alloys, the only way then to effect ajoint is to use a highly corrosive flux such as one of the chlorides (20% HCL works quite well), which then raises the problem of how to clean off the corrosive residues to avoid later damage to the assembly. If many components are involved, this becomes an impossible situation and extremely costly in rework.

When solderability testing indicates that components or PWBs are not to requirements, there are two actions that can be taken. First, reject and return to the vendor. Second, rework the parts and at the same time notify the vendor of the results of the tests and make absolutely certain that he understands the seriousness of the complaint and that he must take immediate actions to rectify the problem. Charging for the rework is one way of making the point quite clear. In this stage of technology there is no possible excuse for supplying parts of anything but excellent solderability.

The one thing that must never be done is to use the parts and hope that by "fiddling" or "fine tuning" the soldering process by changing fluxes or adjusting the soldering machine the product will somehow work out right. IT WILL NOT.

There is one pressure that all ME and QC personnel must be prepared to resist. The product has to be shipped, no other parts can be obtained in time, and the cry goes up, "We have to use them." For this reason it is absolutely necessary to make the entire organization totally aware of the probable results of doing this. On a more positive note it is necessary to be prepared to hot tin dip any components that will not pass the solderability test, but that have to be used.

There are on the market several machines designed to be used for the tinning of components (Fig. 7.13). They generally consist of some form of conveyor that will move the parts either singly or in bulk, mounted on suitable tooling, through a flux and solder dipping station. In some systems the parts are automatically rotated so that both leads on an axial leaded component are tinned in one operation. For some lots manual dipping is quite satisfactory, and in fact is much faster than is sometimes suggested.

The advantage of component tinning of course is that a much more active flux can be used, as the individual components can be much more thoroughly cleaned than when mounted on the PWB. Small components are conven-

Figure 7.13. A typical component lead tinning machine. Courtesy Electrovert Ltd.

iently cleaned by loading into a sieve, after tinning using a very active aqueous flux, and washing under the hot water tap. Aqueous fluxes are extremely useful for this purpose, as their greater activity is usually necessary to ensure complete solder coverage, while they are easily removed when the tinning is completed. Of course any nonsealed component must be handled with care to avoid any retention of flux which may cause later problems, or damage from the cleaning. Parts must be completely dry before being repackaged or sent for assembly.

Poor solderability of PWBs is a much more difficult problem. If the boards have been solder plated and reflowed and exhibit poor solderability, the chances are that the problem lies in the copper plating, and there is nothing that can be done. It is possible that hot air leveling may save scrapping them entirely, but this is usually beyond the capability of the assembly house. Similarly, any plating finish cannot be reworked except by the board vendor.

With copper finished boards the problem may be with the plating, or with the copper foil, depending on the process used to produce the board. In these cases, again, there is little that can be done except to reject and return to the vendor. However, if the boards have been in stock for some time, the problem may be nothing more than an excessive buildup of copper oxide, which can be removed chemically. A dip into a dilute solution of hydrochloric acid (about 5%) followed by a wash in clean water and drying will often be enough to restore the solderability of the boards. Alternatively, there are several proprietary copper cleaning or brightening solutions on the market and sold to the PWB fabricators. It is useful to try these, and to have one available for emergency use.

As previously discussed, if the boards have a protective lacquer finish it may be necessary to wash this off with a suitable solvent prior to soldering, if the solderability testing shows that they are defective. Remember that once the lacquer is removed the boards must be treated as plain copper finished PWBs.

All these methods of restoring the solderability of the parts must be considered as emergency tactics only. Some companies have set up component tinning as a regular part of their assembly process; some tin everything to avoid the need to control the shelf life. This is not the right way to arrive at the most economical production, and zero defect soldering is as much concerned with the economics of making good joints as with the quality of the final product. The component and board vendors have a duty to provide parts that are perfectly solderable, and that will retain that solderability for a reasonable period of time. They can do this most economically, but will not do so unless you, the purchaser, demand it and make them aware of the importance to your operation. When solderability problems arise therefore, the very first thing to do is to contact the supplier. Discuss the problem with him. Get him into your plant to see the seriousness of the problem and its effect on the cost of your process. Work with him to provide parts with excellent solderability. To summarize:

Set up a properly organized program of solderability testing.

Use good statistical methods and plain common sense to keep testing to an absolute minimum.

NEVER USE PARTS THAT WILL NOT PASS THE SOLDERABILITY TEST.

If the parts must be used, tin them before assembly

Copper boards can be chemically cleaned to restore solderability.

Most other PWB finishes can only be reworked by the vendor.

Make absolutely sure that solderability requirements are included in all component and board purchase orders.

The finest solderable finish is hot solder tinning.

Work with the vendor to achieve and maintain excellence in solderability.

Solderability is the most important single factor in achieving excellence in soldering. Without this everything else is wasted effort.

TOTAL PROCESS CONTROL

The major components of the zero defect soldering program have now been reviewed in detail:

Understanding the fundamental principals of soldering
Learning the physics and chemistry of making a good soldered joint
Learning about the basic mechanisms of the solder machine
Setting up the operating parameters
Understanding the importance of solderability
Controlling the process

These many skills and areas of expertise now have to be assembled into one total organization to control all these factors, as only then will it be possible to ensure that all the soldered joints produced are of perfect quality and made at the lowest cost.

This is essentially a matter of setting up the correct information flow, allocating responsibilities, and assigning authority to make certain decisions. The problems cease to be those of technology and engineering but move to the much more difficult areas of people and attitudes. It may be considered that this is out of place in a technical book and that this is not the place for discussing organizational matters. However, after completing the technical direction of setting up a system for zero defect soldering, this part of the procedure is as much a reason for success or failure as anything previously discussed. In fact more soldering systems fail to produce the expected results because of organizational or management failures than those that fail for technical reasons. When lecturing on this subject, it is not uncommon to find that all of the problems are known, the cures are understood and available, but because of the way that the company is organized they cannot be applied. As an example, test or field failures are not reported back to the manufacturing engineer responsible for the soldering process. The solder machine operator is responsible to the production department and is free to adjust the machine parameters as he thinks is correct, or to keep up with production variations. Boards or components are rejected for inadequate solderability but management demands that they be used to maintain the schedule. The solder machine is seen as the only source of defective joints and is constantly readjusted in an effort to correct the problem. The examples are too numerous to mention, and anyone with experience in this area will understand the difficulties that this can generate.

This undisciplined method of operating cannot be tolerated. As with all true control systems, there must be feedback from the results of the process to enable changes to be made to compensate for variations. In zero defect soldering this feedback comes from three areas and each covers a different time span. These three areas are

Solder joint inspection immediately after soldering
Electrical test where any failures due to soldering problems must be reported
Field failure reports of defect due to faulty soldering

It can be seen that the solder joint inspection will produce results that can be used to take immediate actions. If the operator finds that there are some missed joints on the board with no solder, the machine can be immediately shut down and the cause of the problem found. Very few faulty assemblies will have been processed, and those that have can be collected and reworked. The results from the test area, however, will come much later after the boards have been soldered, and quite a large volume may have been incorrectly processed. Finally, the field failures can only be used as statistical data, because of the time from the actual processing of the assemblies, and will be quite useless unless precise data of the processing parameters have been collected and maintained. However, these are all of the utmost importance in maintaining a long-term control of the process and keeping confidence in the system.

In addition to the problem of working with historical data, it must be remembered that the percentage failure rate that is being controlled is extremely tiny, fewer than 100 p.p.m., and indeed anything greater than this is a sign that control has been completely lost, and the process must be shut down until order is restored.

The main and immediate controls are, therefore, those generated by the processing parameters themselves: the solder contamination control, flux density, conveyor speed, and so on. These are the only tools with which the day-to-day operation of the process can be assured, and this explains why the maintenance of these records has been emphasized so thoroughly in the preceding chapters. Each of these processing functions contains its own feedback loop to ensure control, and the data produced enable accurate, detailed records to be maintained (Fig. 7.14).

At the center of the zero defect soldering process is the *solder process controller* (SPC). He receives all the information regarding the operation of the process; he alone has the authority to change processing parameters. The data from the field and test failures are routed to him, and also details of any variations in the product design. Purchasing and QC send him reports of any component changes, and from the assembly floor he receives continuous data on the performance of the soldering system from the solder joint inspector. His method of control of the process is based on two very simple truths, which are usually ignored or forgotten:

> Once the soldering parameters are correctly determined, the results will always be the same provided that nothing is changed.
>
> Any change in results indicates that something has changed in the materials, or in the processing parameters; find the variation and effect the cure.

Therefore, his task is to control all possible variations to achieve a continuously acceptable level of quality as judged by the solder joint inspector. Remember, however, that the inspector is only sampling the process and all

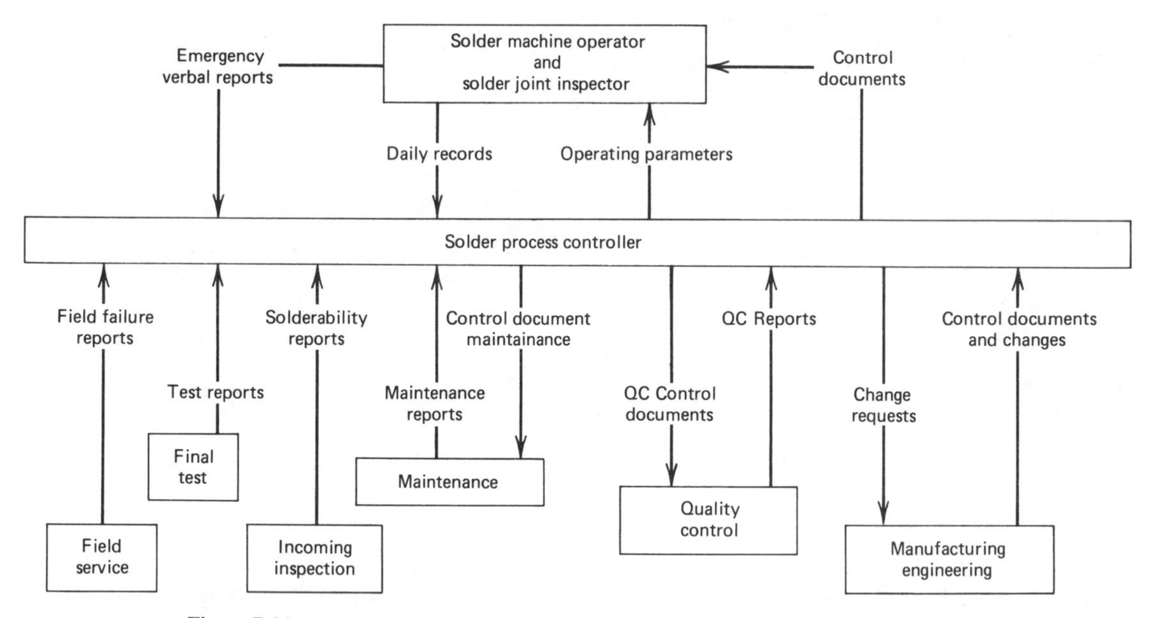

Figure 7.14. An organization chart of a typical zero defect soldering control system.

the initial parameters must have been set up to achieve this level without any touch-up. He is not there to initiate change or improvements in the soldering, merely to monitor results.

Any apparent increase in the number of rejected joints must be reported immediately to the SPC, who will then determine the course of action to be taken. The primary feedback loop for quality is thus dependent on the skills and motivation of the inspector, and this person must be chosen and trained with the greatest care.

It has been found advantageous to rotate the solder joint inspector and the machine operator every hour to avoid the fatigue and boredom which are the enemy of visual inspection. They must of course be trained in both tasks. The rotation of jobs also ensures that both have a very practical understanding of the process and encourages cooperation in producing an excellent product and controlling the process. They must be free from any other duties that may divert from the main task. The inspector, for example, must not be expected to load or unload the machine. He must concentrate on measuring soldering performance.

The objective is to maintain solder joint reject levels at zero, which in all reality is probably not possible. But this must be the aim, and unless the failure rate is within a few per hundred thousand, the process is not reliable, and the cost of rework unacceptable. This means that the inspector is on the lookout for an occasional failed joint only, and it is extremely difficult for him to know whether the failure rate has moved from say 0.001 to 0.003%, that is, a trebling of defects. This is where the test failure reporting becomes important. These data are the first numerical measure of solder joint failure, but it may be several hours or even days from the actual time of soldering. This is why accurate and detailed record keeping is such an important part of the process. When an increase of solder joint rejects is reported, the worst thing that can be done is to immediately rush off and try to "tweak" the process or the solder machine in the hope of effecting a cure. Remember the second rule: find the variation to effect the cure.

When a report of a change in results is rcported to the SPC he will then review his records for the time when the boards were soldered. Was the solder at the right temperature? What was the flux density? Were any particular components involved, and, if so, what was the incoming inspection report regarding solderability? The SPC continues to examine all the possibilities. It may be that sometimes the fault is never found. Some batch of components may have had a few with dirty leads, or an operator may have inadvertently contaminated a PWB by handling. But at least the process will not have been changed because of someone's hunch as to the problem, which inevitably leads to a worse condition. Usually the fault can be traced to a particular item or action, and although the assemblies that have been soldered will have to be reworked, the processing parameters can be changed to avoid the reoccurrence, or better board tooling developed, and so on. Similarly the reports from the field can be related back to the conditions that existed when the assemblies were soldered. Perhaps the failure rate goes up for the day or two before the flux is dumped, which indicates that it should be changed more frequently, or the cleanliness of the assemblies improved. With all of the statistics available there is a chart to steer by. Without adequate data there is nothing to work with, and the process will be hopelessly lost. Intuition is no alternative for hard facts.

So it can be seen that correctly set up, with the correct organization, the system is not only self-controlling but is also self-improving. There should be a steady reduction in solder joint failure once the concept of zero defect soldering is introduced and the statistics begin to be collected. In addition to improving the solder joint quality, the data will be invaluable to purchasing in the assessment of vendors, and to engineering for determining the effectiveness of design features in obtaining reliable joints. Even the assembly operation can use the information to find out if the handling methods are detrimental to the soldering process.

The SPC must be skilled in the collection and interpretation of statistics, and must be given the authority to control totally all the aspects of the soldering process. He, therefore, requires the following inputs:

A report on the soldering results from the joint inspector at least once every shift, once every hour if the volume of product is high; the inspector must shut down the solder machine and report immediately if a major problem occurs

A report on the failures found in test related to solder joint quality, normally once every shift, but immediately if any major problem is found; these data must include adequate information so that they can be related back to a particular lot or batch number

A daily report on the assemblies processed, including lot and serial numbers, so that they can later be referred to individual equipments

Time and shift of soldering related to each lot

Daily report on the results of solderability testing of incoming materials, with adequate identification to be able to refer these test results to particular product lots

Daily report of solder and flux control: analyses, dumping, refilling, and so on

Maintenance reports of every service or repair to the solder machine or ancillary equipments

A QC report on all the various functions that they are required to monitor in the soldering process; QC to notify the SPC immediately any part of the process is found out of control

Any changes to the personnel concerned with the soldering process must be reported to the controller, and he must be given the opportunity to approve any replacements

A file of the processing parameters of all the products to be soldered

It can be seen that in some ways the SPC is acting as a project manager. He has ultimate control of the project and yet no one actually works for him. He brings together many people with different disciplines and skills to achieve a single objective.

The SPC does not actually produce any of the technical instructions himself, for example, manufacturing engineering will be responsible for setting the processing parameters and the maintenance requirements. These will be set out in formal documents approved and signed by the SPC. They will be strictly adhered to, and compliance must be monitored by QC. Once set up, however, they must not be changed in any way without the approval of the controller, who in turn will consult with the appropriate engineering department to determine the correct action to be taken. The SPC will have one advantage over the various engineering groups, in that he will have the data and the view of all of the various activities that can impact on the soldering process.

How does all this reflect on the cost of the process? Although now formally recognized with a title, and clearly defined responsibilities and authority, the SPC is only taking over the tasks that should have been carried out in any efficient mass soldering operation. There will be more data collection than is usually the case, but again this is only formalizing and organizing activities that are probably being carried out in other ways, and that should have been happening in any well-controlled system. Probably there will be more cost seen in the procuring of components and the testing of their solderability. However, these costs will be offset by the practical elimination of the touch-up and solder joint inspection department, a reduction in the cost of final test, and above all in the higher quality product that leaves the factory, and the reduction in the cost of field service. Most important of all will be the reputation that the high-quality trouble-free product will bring to the company. This reputation is priceless, and, in these days of extreme pressure of competition in the electronics industry, is necessary just to survive.

TRAINING

From time to time throughout this book there have been references to the need for training, and this is an important factor in the drive toward zero defect soldering. Traditionally the solder machine operator has been chosen from the personnel on the assembly floor, frequently on the basis that the operator who is not very good at assembly can be used on the solder machine. He is then given a day or two to watch someone else do the job, and then left to carry on as best as he can. This is just not good enough. It is not fair to the operator, and will not produce the required results from the process. The present day solder machine is a sophisticated piece of machinery, and the operator must be properly trained to use it effectively. It is suggested that the operator should also carry out the task of solder joint inspection, as this is the only way that he can ensure that the process for which he is responsible is functioning correctly.

He must, therefore, be trained in both the areas of machine operating and joint inspection. It is not sufficient that he is taught these skills in a parrot fashion. The operator will meet with many unforeseen conditions due to the many variables in the overall process, no matter how well it is controlled, and he must be able to arrive at logical conclusions, based on the results that he sees. As a bare minimum he must have a clear understanding of the following areas of soldering:

Solder wetting of the base metal and the formation of intermetallic compounds

The action of fluxes, why they are necessary, and how they work

The composition of solder, and the effects of impurities on the solder joint

The importance of solderability to the soldering process

The effects on the soldered joint of the various adjustments of the solder machine

The thermal considerations of making a good soldered joint

The appearance of a perfect soldered joint

Faulty joints, their appearances, the correct descriptive terms for them, and their probable causes

Joint failure that requires immediate shutting down of the machine

The importance of accurate and consistent recording of data

The importance of following instructions precisely

These skills and knowledge are not acquired quickly, and not by haphazard training. A properly organized program must be set up, to include the theory and on the job training, and until the various departments concerned with the soldering process are satisfied that the operator is competent, and has been approved by the SPC, he must not be allowed to take control of the soldering machine.

It is often difficult to find competent instructors in-house with either the time or the skills to be able to carry out training to the level required, and consideration should be given to using a local technical college or an outside consultant specializing in this field. The SPC will in most cases also require some additional training, and should take the same course as the machine operator. It is unlikely that someone can be found with the engineering and statistical skills that this job requires, and deficient areas must be made good with adequate training. Bear in mind that this individual is not so much concerned with the actual engineering of the soldering process, which is the task of manufacturing engineering, but rather with the collecting and interpreting of data, acting as a central control, and pulling together as a team all those departments that affect the soldering results. He, therefore, also needs to be a good manager, capable of motivating all those concerned, to carry out their various tasks and to bring the whole to a successful conclusion. He therefore must be skilled in the following:

Statistical analysis of results

The technical background of the soldering process

Managerial and motivational ability

Clear logical thinking

Training must not stop at the people directly involved in the soldering process, but must extend to those who are on the fringes of the operation, but whose efforts can affect the results, or even make the objective of zero defect soldering impossible to achieve.

Engineering must be trained in the design techniques that avoid soldering problems.

Purchasing must be informed of the importance of the solderability of the purchased parts.

The assembly floor must understand the importance of good housekeeping, cleanliness, and the correct handling of the parts.

The stock room has to be trained in the correct storage and handling techniques.

Receiving inspection must be instructed of the importance of the solderability testing.

Everyone has to be made to understand the importance of accurate and timely reporting.

This training is the responsibility of the SPC. There will doubtless be some opposition to the idea that the soldering process is such an important part of the manufacturing operation that other departments should incur additional expense and effort to improve the soldering results. This is where the SPC must have the total backing of top management, and must prepare the cost data to show the importance of zero defect soldering to the profitability of his company.

In summary then zero defect soldering requires these things.

Trained motivated people

Excellent data collection and interpretation

Accurate, detailed instructions

Well-maintained, up to date equipment

Total process control

Perfection as the only goal

HAND SOLDERING TRAINING

The subject matter of this book is primarily concerned with machine or mass soldering. However, it is inevitable that from time to time hand soldering is necessary for various reasons. Once the soldering process is properly controlled it is not unusual to find that hand soldering is the major source of defective joints. Therefore this section is intended to assist in developing the skills and training necessary to eliminate these defects.

Training operators in hand soldering falls into three main sections:

Instruction in the basic theory of soldering

Instruction in the manual skills necessary to make a sound joint

Gaining experience in a controlled manner under the surveillance of an instructor

In addition there must be provisions made for reviewing the operators at regular intervals and retraining them as necessary. This means that every company must have a formal in-house on-going hand soldering training program. The first section need not occupy more than a few hours but the operators must understand the fundamental factors involved in making a reliable joint. This should be taught in a simple practical form, covering such items as wetting, solderability, the formation of the intermetallic, heat flow, and so on. The first chapters of this book will provide ample information to set up this part of the program.

The second section introduces the operator to the soldering iron and should include instruction in the use and maintenance of the iron, and some practical soldering of joints. The objective is to give the operator sufficient practical instruction that the manual requirements are clearly understood. This section will occupy only a few hours, but then the operator will need time to develop the skills that only experience can provide.

Ideally the trainee should now begin to solder simple routine work under the close scrutiny of the instructor. As experience is gained, less and less supervision will be required until the trainee can proceed to more complex work. When the instructor is satisfied that the operator has all the necessary skills and understanding a practical test should be arranged, and upon successful completion the operator should be approved to carry out any hand soldering operations.

All operators must be checked from time to time to be sure that bad habits have not been developed. This is not to infer that operators deliberately use incorrect procedures but rather that we are all human and frequently only an outsider can see precisely our working methods.

Some of the most common bad habits are

Soldering too quickly so that the joint does not have time to wet properly.

Applying too much solder in an attempt to make the joint look "rounded."

Flicking the iron upwards at the end of the soldering.

Stroking the iron up over the lead at the end of the soldering.

Pressing down on the iron.

The review procedure need be no more than a close inspection of the operators at work on the line. Once a year a more stringent test should be arranged which must be passed before any individual can continue as a hand soldering operator. A formal certification program for hand soldering operators has been found to be particularly effective, not only in controlling the process but in demonstrating the importance attached to hand soldering with respect to the product quality.

The following information has been laid out as a short instructional chapter for the hand soldering operator. Some parts duplicate information found elsewhere in this book but has been modified to relate specifically to hand soldering.

Making the Joint

When a soldered joint is made a tiny portion of the molten tin contained in the solder combines with the surface of the copper in the board and leads to be joined and forms a very strong intermolecular bond. It is this bond which provides the strength of the soldered joint and which is much stronger than the actual solder itself. This combination of tin and copper is called an *intermetallic compound* with the formula of Cu_6Sn_5 and Cu_3Sn depending on the amount of available tin in the joint.

This compound is not only very strong but is also very brittle especially when it is thicker than normal. Under these conditions a joint can crack when subjected to stress such as thermal changes or mechanical forces.

The thickness of the intermetallic compound is largely determined by the length of time that the solder is molten and therefore soldering must always be carried out as quickly as possible.

Flux

In order for the molten tin to form this intermolecular bond with the materials to be soldered, they must be perfectly clean and free from metal oxides. Flux is necessary to remove the metal oxides which are always present and prevent them reforming while the joint is being made.

The flux does this once it has been melted and heated by the soldering iron. The solder will not bond or *wet* to the metals until they are coated with the flux and it has been heated to a temperature at which it will become chemically active and remove the metal oxides. This temperature is about 260°F (127°C) for rosin based fluxes. However, if these fluxes are heated much above 600°F (316°C) they will char and become useless. The temperature of the soldering iron therefore is very important.

The flux is usually in the form of a solid material which is found running down the center or core of the solder wire. (Hence the name "cored solder.") When the solder wire is applied to the iron, the flux first liquefies and runs out over the metals to be joined, preparing them for the molten solder. Flux is sometimes used in the form of a liquid, which is a solution of the flux, usually in alcohol, and is applied to the joint prior to the application of the iron and solder.

Solder

Solder is a mixture of tin and lead. The solder which is usually used has 63% tin and 37% lead. The molten tin is a very good solvent of other metals and is the active ingredient in making the joint. Solder with this composition or with a composition very near these percentages melts at 361°F (183°C). As the composition moves away from these figures the solder turns into a paste at 361°F (183°C) and only becomes a true liquid at much higher temperatures. Solder with the 63/37 composition is therefore called a eutectic alloy, which means that this material will turn directly from a solid to a liquid when heated.

Solder is usually used in the form of a wire for hand soldering electronics, with the flux as an integral core of the solder. However, this is not always the case and solder can be supplied in the form of a wire with no flux, as a paste, in the form of tubes, sleeves, or washers (known as preforms), with or without a flux coating. Solder can also be used with widely varying percentages of tin and lead and with the addition of silver or other metals to produce solders with very different properties. However, most soldering is carried out with solder as described at the beginning of this section.

The Soldering Iron

The soldering iron is used to heat the joint to be soldered and to melt the solder. It is composed of the following basic parts: the heating element, which is connected to the electrical supply, either directly or through a transformer or a control system; and the bit, which is heated by the element and is usually made of copper, plated with some material which will wet with solder but will not be dissolved quickly by the molten tin in the solder. The plating can be iron, steel, nickel or some more exotic alloy. The end of the bit, which is the actual working surface, is called the tip and has to be wetted with solder to be used to make the joint.

Heat has to flow from the element to the tip and the ultimate idling temperature of the tip is determined by the wattage of the iron, the size and shape of the bit, and the overall design of the complete iron. A simple iron will have a widely varying tip temperature which will depend on the thermal load applied to the tip. For example, at idle the tip temperature of a particular iron may be at 900°F. When soldering a large joint, or several joints immediately one after the other, the tip temperature can quickly fall by several hundred degrees if the heat is pulled from the tip faster than it can be replaced by the element. To use a simple iron therefore the iron design must be selected so that the heat flow from the element to the tip will match the amount of heat removed from the tip during soldering and the tip will remain at a reasonably constant temperature.

With a temperature controlled iron the amount of electrical energy applied to the element is controlled according to the temperature of the tip. As heat is removed from the bit the control system automatically supplies sufficient electrical power to bring the bit back to the controlled temperature. A controlled iron therefore if used within the load capability of the iron and having the correct bit design will maintain a tip temperature which varies only a few degrees.

There are two types of controlled irons:

Temperature limiting
Temperature controlled

While the first type is usually satisfactory for normal factory use, the second, and more expensive iron is preferred for rework or other areas where joints of differing thermal masses have to be soldered.

With all irons the heat has to flow from the element to the tip through the bit and the selection of the correct bit size and shape is important to keep the temperature drop as small as possible. The soldering iron bit should be

As large as possible
As short as possible
Tapered, not turned, if this is possible
Have a tip as wide as the pads to be soldered.

Care of the Iron

While most soldering irons are robust tools they should be handled with care. Do not use the soldering iron as a probe, screwdriver, hammer, or in any other ways that apply mechanical force to the iron. Do not kink or twist the lead, and never wrap it around the iron. Do not pull on the lead or the lead connections. When the iron has to be stored wind the cord into a large loop and do not let it touch the iron when hot.

The bit must be removed at least once a day and the shank wiped clean of scale, flux, or other residues. The iron should be tapped gently to knock out any scale from the socket that retains the bit. When assemblying the bit back in the iron make sure that it is fully pushed down and any set screws completely tightened. If the bit is not regularly cleaned oxides can build up and prevent the flow of heat from the element. In addition with time the scale can prevent the bit from being removed from the iron and then the entire element, bit, and casing have to be replaced.

The iron should always be placed in the proper holder when not in use. The holder is designed to dissipate heat as may be necessary and helps to keep the bit at the correct temperature. The tip must always be tinned with a thick coating of solder before the iron is placed into the holder.

The sponge that is used to clean the tip should be kept wet. This means that it is not sufficient to make the sponge damp; it must be saturated with water. If a finger is pressed into the sponge it should be possible to see water at the bottom of the depression. It is the steam from the water that is intended to clean the bit. If the sponge sticks to the bit or if the sponge is burned then it is not saturated properly with water.

Forming the Joint

In order to make a good joint the metallic parts to be joined must be heated to a temperature above the melting point of the solder. In order to prevent damage to other parts, such as components or the PWB, the heating must be carried out as quickly as possible.

Heat is not transferred from the iron quickly, unless the tip is connected to the materials to be joined by a layer of solder wetting them to the iron tip. Therefore the soldering iron tip is applied in such a way as to touch both the component mounting pad and the lead to be soldered. Cored solder is applied to the iron tip immediately adjacent to the lead and pad. Heat from the iron will melt the flux, which will flow out onto the parts to be joined, removing the oxides and cleaning the surfaces in preparation for being wetted with the solder. Almost immediately the solder will begin to melt and will flow out over the pad and lead (or other parts being soldered), and the joint will be formed. The solder wire is then removed when sufficient solder has been applied to the joint to form a satisfactory fillet, the iron is taken away and the solder solidifies.

All of this takes only a few seconds, in the case of a single-sided board joint about 1.5 to 2 seconds, in the case of a PTH about 2 or 3 seconds. The exact time will depend on the thermal mass of the joint and must be kept as short as possible provided that the time permits the entire joint structure to reach soldering temperature and the solder to completely flow over the entire joint. It is the "wetability" of the materials to be soldered that allows the solder to flow out and form the joint. If the parts or board are not clean or solderable then wetting will not occur and there is no way that a satisfactory joint can be made. Do not attempt to make the joint look good by piling on solder or attempting to move the solder around the joint with the tip of the iron.

If the joint is connected to something with a large thermal mass, such as a large terminal, heavy semiconductor lead, ground or voltage plane, or a solder tag bolted to a metal structure, it may be impossible to make a sound joint without using a very large or high wattage iron. It is not possible to solder every kind of joint with the same size of iron and bit.

When soldering PTH boards the iron and solder are applied to the solder side of the board and the solder is required to flow through the hole and form a fillet on the component side (top) of the board. This means that the entire hole structure and lead must reach soldering temperature and sufficient solder must be applied to produce a top fillet. As the top of the board cannot be seen while soldering it requires some experience to ensure that a top fillet is formed each time and that neither too much nor too little solder is applied. With too little solder there will be no top fillet or it will be inadequate. Too much solder can produce a solder short or the top fillet may be so large as to include the stress relief bend in the component lead, which is not acceptable. Experience permits the operator to develop a rhythm for this task, but until this experience is developed, examine the top of the board from time to time to be sure that the correct fillet is being formed.

The soldering iron should be held very lightly with no pressure applied to it at all. At the end of the joint formation it should be lifted directly upwards from the joint, without flicking or stroking the solder over the lead being soldered. This allows the solder to flow only by the wetting force and the quality of the joint is not hidden. The entire motion of making a joint should be controlled, but very gently, with no force applied to the soldering iron tip except to guide it to the correct position for heating the joint.

Soldering is not magic, it is not an art, but the operators must be skilled and self-controlled in their approach to the task. They must be supplied with parts and boards of excellent solderability and be given the correct tools and facilities so that they can produce the excellent soldering that alone is acceptable.

TROUBLESHOOTING THE PROCESS

When the process is being set up, or when problems occur during the actual soldering operation, it is necessary to be able to identify the probable reason for the variation from the excellence of soldering required.

Always remember that if the process was operating correctly, then the problem is some variation from the original parameters in the materials or the machine. However, if the problem occurs during the initial setting of the parameters, then it will be necessary to make operating changes to arrive at the optimum settings.

There are many tables or charts published which are supposed to offer "cookbook" solutions to the various problems found in the soldering process. Unfortunately, this does not work because there are too many variables. For example, the chart may show that for an inadequate top fillet there are five or six options: slow the conveyor, more preheat, more flux, raise the solder flux height, and so on. This is not only confusing, and rather useless, but encourages the feeling that by sufficient trial and error the optimum result can be obtained. For this reason so much effort has been put into explaining the fundamentals of the process and the detailed workings of the various parts of the soldering machines. Solutions to soldering problems will be found by examining the available results as carefully as possible, and then using the basic theory already learned to arrive at logical answers. The following information, therefore, must be used with caution and understanding, the interpretation depending on the particular set of circumstances in effect at the time.

No Solder (Fig. 7.15)

As previously mentioned, many of the problems found with soldered joints will be caused by poor solderability of the board or the component. If during soldering, the lead is wetted and the board is not, or vice versa, then the flux and solder must have been present during the operation, and the solderability of the unwetted part must be suspect. If this is found to be an isolated instance, the part may have been contaminated during assembly or other handling; if, however, the problem is more widespread, and the same part is involved in every case, then the solderability test records must be checked, and another solderability test run immediately, with the process stopped until the solution is found.

Where the joint is completely free of solder, check again to see if this is an isolated case or if the problem is more common. If it is seen on several boards, and the solderability of the parts is assured, then the solder wave, or the conveyor or pallet is suspect. The suspicion must be that the board does not contact the wave over its entire surface. Pallets that are out of true and that rock on the conveyor can produce this effect. A jerky conveyor is another cause for the board to have skipped joints. The wave may have

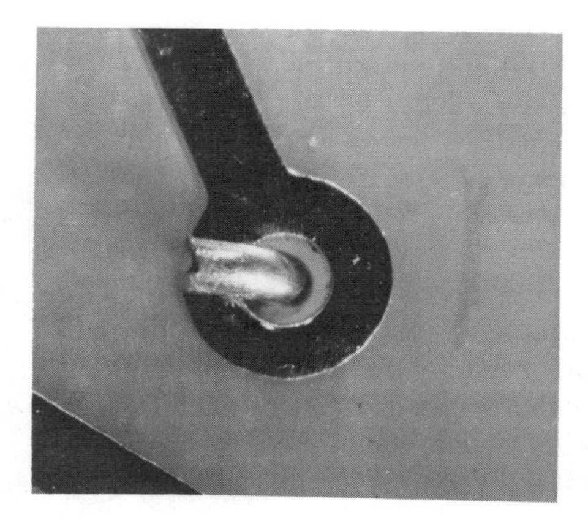

Figure 7.15. A very obvious soldering defect—no solder.

excessive turbulence, or the height may have changed suddenly and momentarily. Review the maintenance records, and if in doubt have the pump and nozzle cleaned. If this problem is found to coincide with the maintenance times, then perhaps the frequency of cleaning must be increased. Skipped joints are almost always related to the accuracy of pallets, conveyor, or solder wave. Incidentally, this shows the need to check the level of the wave, the squareness of the pallets, and any other tooling on a regular basis with the results recorded and on file. Of course, the fluxer can also cause this problem in the same way as the solder wave, but an examination of the joint will usually show if the flux has been applied.

If the boards are not properly held during the actual soldering, there is the possibility that they will warp, and this will raise one area of the assembly from the solder wave, and consequently some joints may not be soldered. This will usually show as a consistent failure, but can be visually checked by watching the board for any movement as it hits the wave. This effect can sometimes be prevented by applying more preheat, and then running the assembly over the wave at a higher speed. A much better solution is to improve the tooling so that the board is held more tightly to prevent any movement.

Excessive Solder

Excessive solder can manifest itself in many ways. Solder shorts and bridges, large globules on the joints hiding the outline of the wires, icicles, and spikes are all signs that too much solder has been left on the assembly (Fig. 7.16). These faults indicate that the conditions were not satisfactory for the excess solder to drain freely following the actual wetting and joining. There can obviously be no doubt as to the solderability of the parts, and the cause must lie in the factors that can affect the flow of the solder, for

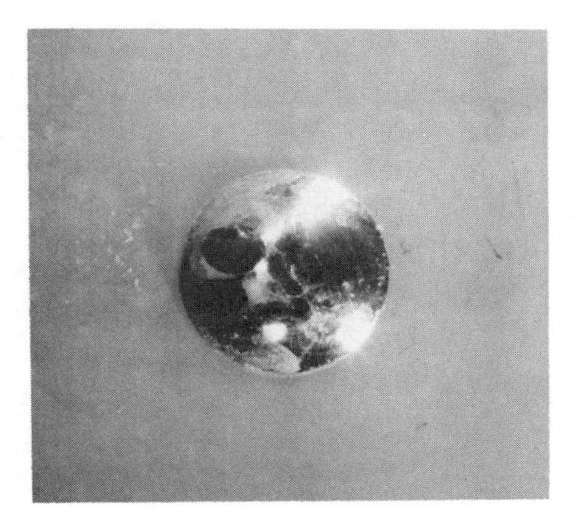

Figure 7.16. This joint is acceptable to most specifications, but the solder is excessive and the following processing parameters should be checked: flux density and solder wave, oil injection.

example, the surface tension of the wave, and the time available for the excess solder to be removed by the force of gravity and the dynamics of the wave design. In the dry wave the flux can have an important effect on the amount of solder left on the joint. Insufficient flux, or flux with too low a solids content, will show up as a dry appearance of the laminate surface after soldering, indicating that there was insufficient flux left after the initial exposure to the wave to maintain a low surface tension during solder drainage. In the wave with oil injection insufficient oil or oil that has not been changed for some time can have the same effect.

The following factors, therefore, should be considered when too much solder is found on the solder joints:

Flux specific gravity
Flux quantity on the assembly
Flux activity; dump and refill if in doubt
Oil injection insufficient
Oil wetting agent no longer active
Solder temperature too low
Conveyor speed too high, not allowing sufficient time for drainage
Conveyor speed too slow, allowing the flux to be destroyed before the board exits from the wave
Preheat too high, flux being broken down by excessive temperature
Solder bath contaminated
Wave shape incorrectly adjusted
Conveyor angle incorrectly set

In most cases of excess solder or bridging, the cause is the incorrect adjustment of the solder wave or insufficient flux in the case of the dry wave, or with an oil injection machine the incorrect adjustment of the oil injection valve. These items should always be checked first. In the case of the dry wave the solder flow pattern is most important and the adjustment of the exit plate must always be made with extreme care.

It can be seen from this list that troubleshooting cannot be a hit or miss affair, and trial and error solutions will not be effective. Only logical step by step tests will provide the answer, based on a clear understanding of the fundamental solution to any problem simple by eliminating many of the possible areas of trouble. Indeed with good control it can be seen that these problems are unlikely to occur once the correct operational parameters are set.

Dewetting (Fig. 7.17)

If the boards and components have guaranteed solderability, wetting problems are limited to the possible inadequacy of the flux or the fluxer. It is just

Figure 7.17. The leads have initially wetted but the solder has withdrawn leaving dewetted areas, a solderability problem with the component leads.

possible that the preheat could be so far out of adjustment that the flux is baked excessively and burned, but the surface temperature of the board can be quickly checked to eliminate this possibility, which is in any case an extremely rare occurrence. Some possible causes then are as follows:

Low flux activity; dump and refill if in any doubt
Incorrect flux density
Preheat incorrect
Conveyor too fast, not allowing time for the flux thinners to evaporate
Too much oil in the wave, so that so much of the solder drains from the joint that is appears to be dewetted
Solder temperature too low

If in doubt recheck the parts for solderability. This is almost always the cause of the problem (Figs. 7.18 and 7.19).

Insufficient Solder (Fig. 7.20)

This problem occurs in two forms, easily identifiable by examining the joint with a 10 power glass. The first form is where the solder has wetted the joint, and has then drained off leaving insufficient solder for many of the current specifications, which demand adequately formed fillets. Under magnification the smooth, shiny layer of solder can be seen. This confirms the solderability of the parts, and shows that the fluxing and soldering operation has been

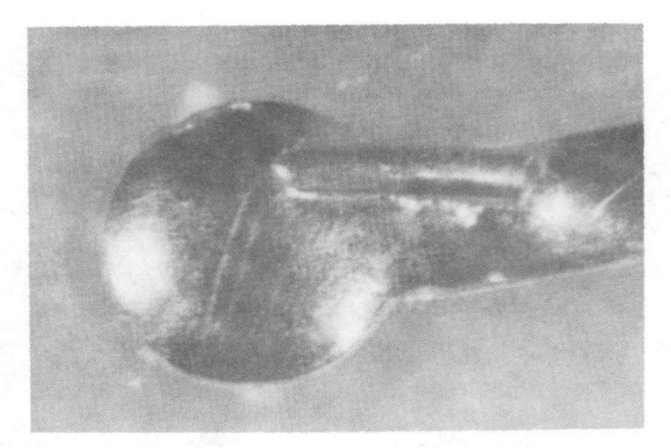

Figure 7.18. The component lead has poor solderability and has not wetted.

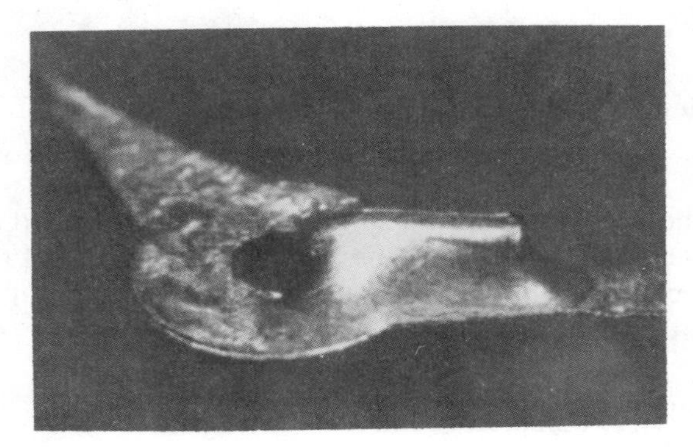

Figure 7.19. The printed wiring board has poor solderability and the solder has dewetted.

performed correctly. What has happened is that the reverse of the conditions discussed in Excessive Solder has occurred. The solder has drained almost completely from the joint, because of the extremely reduced surface tension of the solder wave. This problem is rarely found with the so-called dry wave, and is almost always caused by an excessive supply of oil in the wave, or the use of oil that has excessive wetting agent in it. The only exception to this has been seen where the clearance between the hole and the lead is much greater than normal. In this case, of course, the problem will be seen no matter which wave system is used. The only cure is an engineering change to reduce the hole size.

The second form of this fault occurs when the solder does not wick up through the joint, although the parts are actually touched by the solder and some portion of the joint will be wetted. In this case the problem can be the fluxing action, solderability, or thermal inadequacy. Fluxing action can cause this defect if the flux does not completely coat the joint. The following actions can eliminate these possible problem areas.

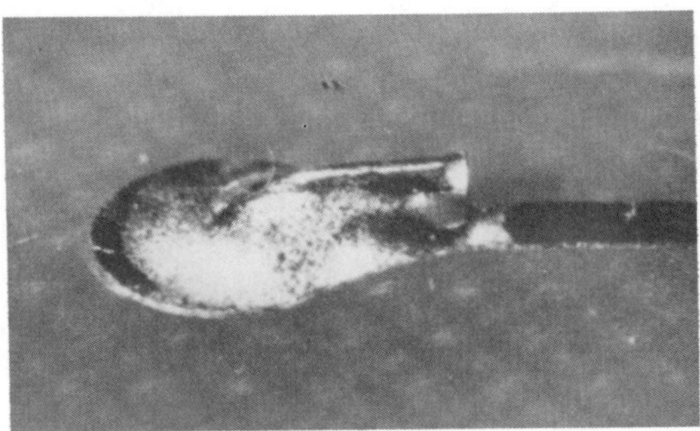

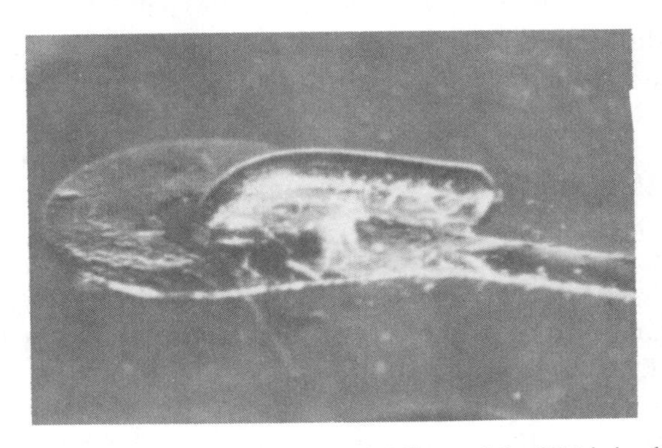

Figure 7.20. Both board and leads show poor solderability and the fillet is inadequate. Note that this is a solderability problem and cannot be solved by readjusting the solder machine.

Check that the fluxer is operating correctly.

Make sure that the flux wets the entire joint; there should be some flux visible on the top of the board after soldering.

Check the flux density.

If there is any doubt as to the activity or cleanliness of the flux, dump and refill.

The question of solderability is obvious as is the action to be taken; however, if part of the joint has wetted it is unlikely that this is the problem. Thermal inadequacy implies that the joint does not get hot enough for the solder to melt and flow, and in this case it does not wick up the entire joint surface; the solder will only rise as far as the temperature of the parts will allow. This problem can be caused by deficiencies in the design of the assembly, or by bad plating in the case of PWBs with plated through holes.

Where the assemblies have components of widely varying thermal capacity mounted close together, the joints of small thermal capacity will be soldered while those of larger thermal capacity will not have reached soldering temperature. In many cases the solder will have heated and joined those parts that are actually in the solder, but the large thermal mass will stop the remainder of the metal in the joint from reaching the temperature at which the solder will melt and fill the joint. The solution is to slow down the conveyor and increase the preheat in an attempt to bring the entire joint up to the necessary temperature. In extreme cases there is a very real possibility that the smaller parts may be overheated, and if the offending components are very large it may prove impossible to increase the temperature sufficiently. This problem is usually seen with such items as heat sinks, bus bars, power diodes and transistors, and similar items where the component itself has to dissipate more than the normal power. In extreme cases it may be necessary to arrange for the component to be mounted in some other way, with low thermal capacity leads into the PWB. Remember that there is a limit to the thermal differences that can be tolerated on any one assembly that is to be machine soldered.

PWBs themselves can cause a problerm through their design, particularly in the case of multilayer boards. Holes passing through ground or voltage planes must have thermal relief or the heat from the joint will be conducted from the joint faster than the solder can supply it, and the plated through hole will be unable to reach soldering temperature past the plane level. This, of course, results in the solder not wicking up the hole, and no fillet is produced on the upper surface of the board.

The chief cause of inadequate fillets on the top of the plated through hole boards, however, is deficiency in the through hole plating. This can consist of thin or missing plating, voids, or cracks. The most troublesome cause is brittle copper, which cracks under the thermal stress of soldering, forming what are often called knee cracks at the interface of the surface copper and the internal hole plating.

This is a particularly insidious failure, as the board will appear perfectly sound, and only a cross section will confirm this particular problem. This failure is often diagnosed as a solderability problem even though the PWB may have successfully passed the edge dip test. It is for this reason that all boards with plated through holes must be given a solder wicking test, and if the solder does not flow over the top pad and completely wet the surface within the specified time, the boards must be rejected, and the test sample sectioned to confirm the diagnosis. There is, of course, the very remote possibility that there may be some other solderability problem, but it is not likely (Fig. 7.21).

Solder Voids, Pin Holes, Blow Holes

Whenever voids are found in a joint it is certain that some action has taken place to displace the solder and form this cavity. Outgassing of the epoxy glass laminate is almost always the cause (Fig. 7.22), with the gasses entering the solder by means of cracks or voids in the through hole plating. Similar failure can occur if the component leads are contaminated, the heat of soldering vaporizing the contaminant. However, this is so rare that it is almost never seen. If the boards are found to be the problem, as is usually the case, baking to drive off any water or volatiles can in some cases effect a cure. The boards should be baked for a maximum of 6 hr at 220°F (105°C). After baking they must be assembled and soldered within 12 hours or the baking will have to be repeated. This is because the boards will pick up

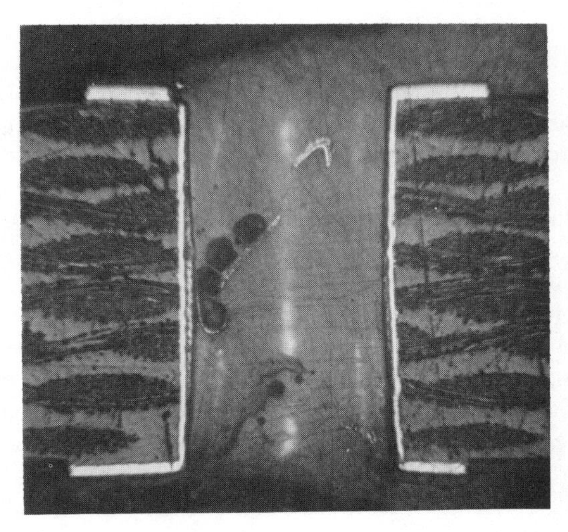

Figure 7.21. The plating in this hole has broken at the knee. There may be electrical failure if it is used.

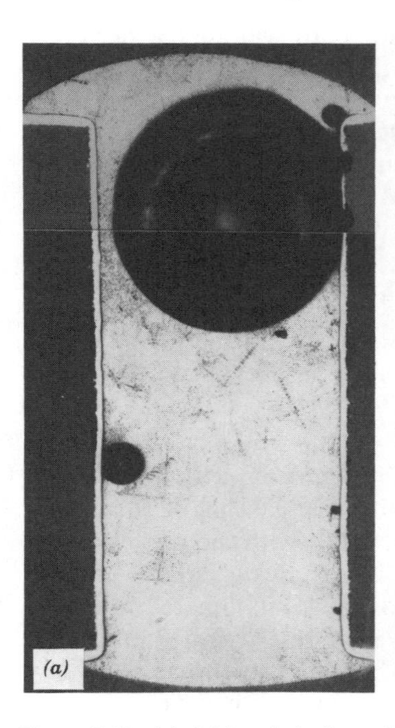

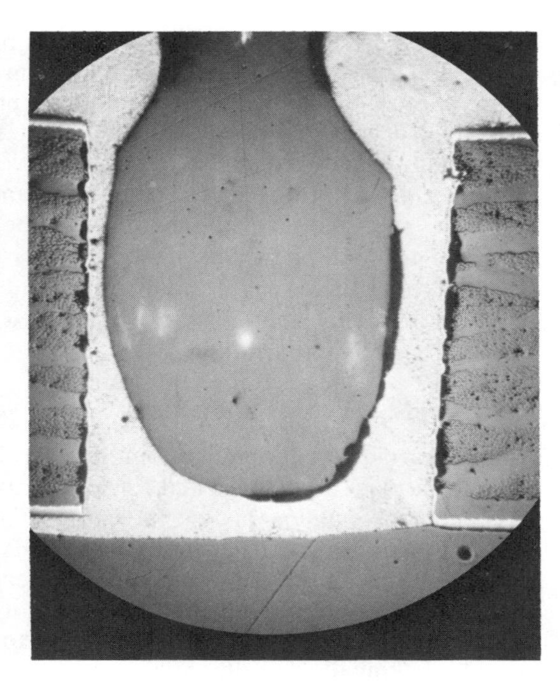

Figure 7.22. (*a*) A blow hole formed because of outgassing from the epoxy glass. The fundamental cause is poor drilling of the hole resulting in breaks in the copper plating. (*b*) In this picture of a blow hole note the cavities behind the plating.

moisture from the atmosphere. These figures must not be considered as absolute. They depend very much on weather conditions, storage methods, and the quality of the laminate. These parameters, however, have worked well in a typical hot, humid New England summer.

In practice blow holes (Fig. 7.22) and pin holes are invariably caused by defective printed wiring boards. The two problems are similar but their cause and effect on joint reliability are quite different. Blow holes originate from poorly controlled hole drilling followed by inadequate through hole cleaning or plating. As a result plating solutions become trapped behind the plating and are quickly absorbed into the epoxy glass matrix. When the hole is soldered the heat from the molten solder generates steam from the absorbed materials, which breaks out through the thin and incomplete plating and forces the solder from the hole. It is not possible to rework blow holes and guarantee a reliable joint. Even if the visible blow holes can be filled it is probable that the barrel of the hole will contain large voids and in some cases may be completely empty with only caps of solder forming fillets on each side of the board. When blow holes are present they will usually be found all over the board because they are caused by a deficiency in the fabrication process. Pin holes are also caused by processing problems but are less serious and rarely affect the reliability of the joint. In plating a glass whisker or an air bubble or other contamination can cause a tiny hole in the plating. At soldering any moisture in the laminate will turn to steam, vent out through the small void, and cause tiny bubbles in the soldered joint. Pin holes are generally scattered randomly across the board in small numbers. They rarely pose any threat to the quality of the joint. It is unfortunate that so many specifications require that they be reworked as the very act of filling the holes only reduces the reliability of an already reliable joint.

It must be remembered that the typical blow hole or pin hole is only seen when the solder freezes as the steam leaves the solder surface. There will inevitably be more invisible cavities in the solder joints. In the case of blow holes, where the volume of steam can be large, if the steam escapes while the solder is still molten a depressed fillet can result. Blow holes therefore can appear in many guises depending on the state of the solder when the steam is ejected.

It is sometimes difficult to differentiate between the two defects. Many specifications define a pin hole as a void in the solder fillet where the bottom of the void is visible and therefore does not require to be reworked. This of course is incorrect as explained above. The best method of identifying the defect is to look at the number of voids visible on the board. In addition it will be found almost impossible to rework any blow hole as the joint will continue to outgas as soon as the solder once more becomes molten. Baking the board as mentioned in the first paragraph of this section is effective in eliminating pin holes. It rarely has any effect on blow holes.

Disturbed Joints

Disturbed joints are joints that have been subjected to some movement as the solder solidifies. The joint will have cracks or fractures where the solder moved, or in less extreme cases there will be lines in the otherwise smooth solder surface. It is sometimes difficult to distinguish between this fault and the next one to be described, frosty joints, as the overall appearance can be similar in some cases. Generally, however, not all joints will be disturbed, and a careful examination with a magnifier will show some joints that appear to be normal. The cause is invariably a jerky conveyor, or an out of true pallet that can rock after leaving the solder wave. In some cases it may be caused by just not allowing sufficient time for the assembly to cool before handling. This effect is not often seen when a eutectic solder is used, because it goes directly from the liquid to the solid phase. Other alloys with a fairly wide pasty range are much more likely to have this problem (Fig. 7.23).

Frosty Joints

The term "frosty joint" covers several forms of rejected but usually reliable joints (Fig. 7.24). The fact that the joint is not shiny and smooth can mean

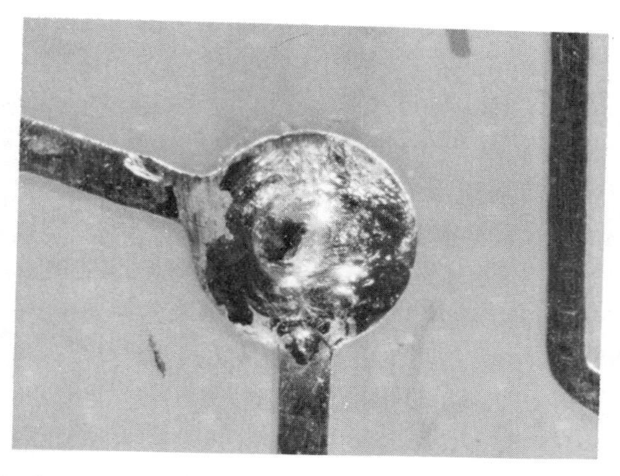

Figure 7.23. The lead moved in this joint as the solder was solidifying, resulting in a ''disturbed'' joint with many microcracks in the fillet.

that it is likely to fail, but the problem is usually only cosmetic in nature and the joint will perform perfectly satisfactorily. Some companies deliberately use a solder with a composition that produces dull joints with the idea of making joint inspection less fatiguing to the inspector. The solder may have a lot of antimony, which causes it to be dull but sound, or a lot of copper that makes it dull, with a remote possibility that it may be more likely to crack under stress. Indeed, there is little or no evidence to suggest that a frosty joint is any less reliable than one that is bright and shiny. For this reason it is

Figure 7.24. Frosty solder as seen (*a*) on a section of board, (*b*) on a single joint, (*c*) on a via hole and circuit trace.

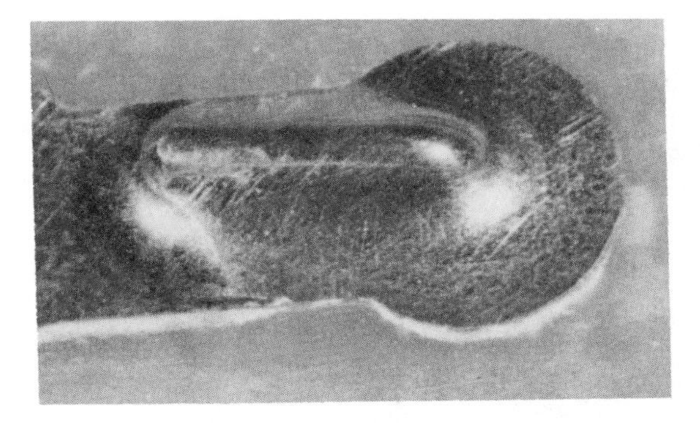

Figure 7.24. *(continued)*

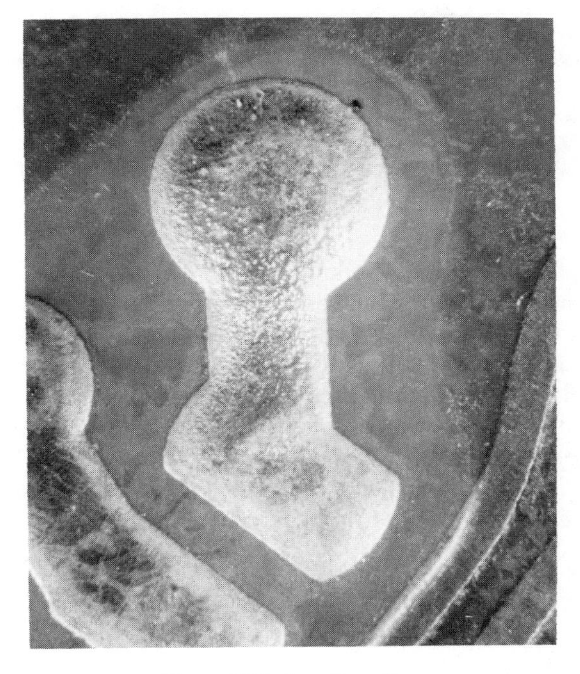

Figure 7.24. *(continued)*

generally considered that frosty joints are rejectable for cosmetic reasons only; it is indeed a reflection on current specifications that so many joints are reworked and rendered less reliable only to satisfy a cosmetic requirement.

Frosty joints should not be reworked. If the joint is gritty or bumpy then the joint may possibly contain a dendritic growth caused by solder contamination. A pure eutectic solder (63% tin, 37% lead) will always produce a bright, shiny joint. This is a very good reason for using a clean eutectic solder, as the onset of frosty joints will probably be the first indication of

Figure 7.25. This is a perfect joint. The solder is smooth, the lead and component mounting pad are completely wetted, and the outline of the lead is easily seen.

contamination in the bath and acts as an early warning that the process control of the solder bath requires attention.

The Logical Approach

The troubleshooting tips given above are those that have been found by experience in the field. They are by no means complete, new problems will always arise, and other solutions will be found; with a thorough knowledge of the basic theory of soldering there will be nothing that cannot be resolved by attacking it in a logical manner. To summarize problem solving the following simple rules should be followed:

Poor or no wetting	Look at fluxing and the cleanliness of the parts
Inadequate solder	Poor wetting as above, or too much oil in the wave
Too much solder	Incorrect solder wave setup, inadequate oil injection
No top fillets	Poor through hole plating
Voids, pin holes, blow holes	Poorly fabricated boards

When setting up the machine to arrive at the correct parameters, make small adjustments, and then run sufficient boards to be sure that the results are consistent. Make only one adjustment at a time. Do not be in a hurry or try to make shortcuts. Keep meticulous records. When a change does not produce the expected improvement it must always be possible to go back to the original setting.

ANYTHING OTHER THAN A PERFECT JOINT MUST BE SUSPECT (Fig. 7.25).

Chapter 8

Special Soldering Systems

Throughout this book the emphasis has so far been on the general principles of soldering, and the machines described have been those that are found in almost any electronics assembly plant. While practically every machine has a fluxing, preheat, and soldering station, there are some machines that do not follow the general pattern, and either perform additional functions or use the standard parts of the process in a different manner.

Some of these special machines may be of interest only as a curiosity, whereas some may be found to be the solution to a particular assembly problem. At least, these brief descriptions may generate ideas that will lead to more efficient uses for the machine soldering process. The larger soldering machines are built to the customer's requirements from a series of standard modules and therefore there is usually no difficulty in including in the design any unusual requirements.

The first part of this chapter will concentrate on the smaller variations; the complete systems will be reviewed later.

The "job shop" has to handle many different types of assemblies, and therefore may require to use different fluxer for different customers. One company solved this problem by having two fluxers built into the machine. The pallets were coded and operated microswitches at the entrance to the conveyor. These switches in turn controlled air cylinders, which raised or lowered one of two fluxers containing different fluxes. One was filled with a rosin based flux for assemblies that could not be adequately cleaned, while the other was filled with an aqueous flux for those boards that were later to be water cleaned. The coding signal also signaled the conveyor to divert the soldered assemblies to the correct cleaning station.

In a different situation the shop was soldering very large numbers of boards, which were similar in design, but were of different widths. As they were not large boards, full use was made of the capability of the machine by installing two conveyors on it. Each conveyor was fitted with its own drive motor and speed controller, and each was of the palletless type and adjustable in width. This modification doubled the throughput of the machine at the cost of a second conveyor.

In another instance of designing for mass production, all the boards that were to be processed had edge connectors of the same cross section, although they varied in length. The machine was fitted with a special conveyor that transferred the assemblies by the connector housing, while the other side of the board was located by a support bar. Also fitted were adjustable masks on the fluxer and solder wave so that the gold fingers on all the boards were protected automatically.

These are only a few examples of the possible modifications. Look at your own needs and decide if some such variation can improve your efficiency.

SOLDER-CUT-SOLDER (Fig. 8.1)

In most assembly operations the components are clinched to retain them during soldering. Sometimes they are preformed and retained by the shaping of the leads. Occasionally they are formed and cut and retained by other means. Skin packs and adhesives are both used; sometimes, the components are just allowed to sit on the board held only by gravity. In this case careful handling is necessary to avoid their becoming dislodged.

None of these methods of retention is without some negative effects. Shrink packaging works well where the components are approximately the same height; if they are not, then small parts located adjacent to tall parts are not properly held and can be dislodged. The main problem with this system, even when the components are the same height, is the difficulty in removing the plastic after soldering. The hot component leads melt the plastic, and effectively bond the plastic to the assembly. Petroleum jelly has been used to coat the leads and prevent the adhesion of the plastic, but this is messy, difficult to remove, and does not always prove effective.

Cloth bags filled with shot are also sometimes used to retain components, but again they are not totally effective, and some parts are always found to be raised up and require rework. Preforming the components and allowing them to sit free on the board is a simple and efficient method of assembling the parts, if the handling problems can be overcome, but requires special attention during the soldering to avoid the lifting of the components. The solder wave will tend to float the parts up out of the holes until the leads wet, when the collar of solder around them will pull the component back down onto the board. This requires excellent solderability, and fast wetting if the board is not to have many of the smaller parts moved or lifted, and even then it is inevitable that some will be disturbed. In addition, all these methods require additional handling of the parts, once to form and trim, and once to assemble onto the board. These conditions especially apply to the assembly shop where for various reasons the use of component insertion machines is not a pratical or economical proposition. The use of the solder-cut-solder system is one method of improving this assembly productivity while retaining the flexibility of the manual assembly line.

In this process the components are assembled with no cutting or forming of the leads, or forming only. The long leads make the components easier to handle, offer greater leverage if they have to be bent, and, of course, the leads retain the components much more firmly once they are inserted into the holes in the PWB. There is, therefore, not so much concern about handling the assemblies before soldering, or for the lifting of components when they meet the liquid metal.

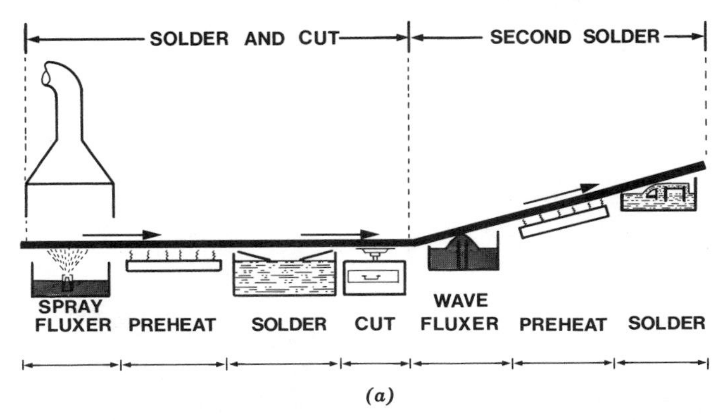

Figure 8.1. (*a*) A schematic diagram of a solder–cut–solder machine.

Figure 8.1. (*b*) A typical rotary blade cutting station. Courtesy Zevatron GMBH.

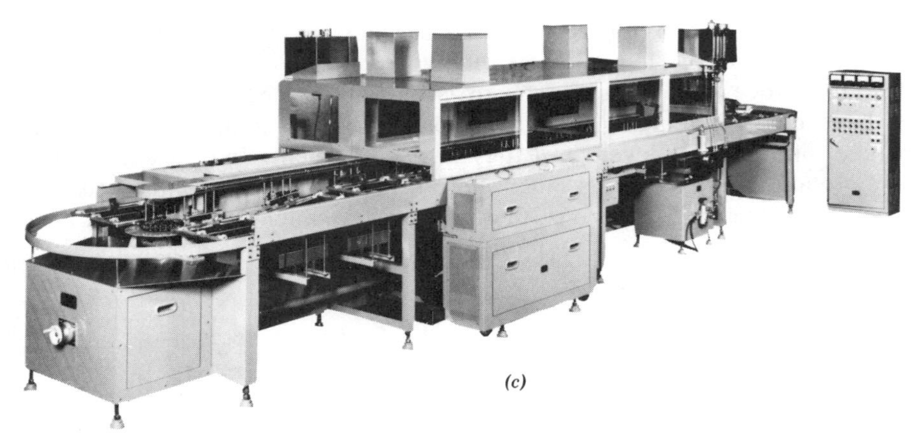

Figure 8.1. (*c*) A complete solder–cut–solder system. Courtesy Tamura Seisakusho Co. Ltd.

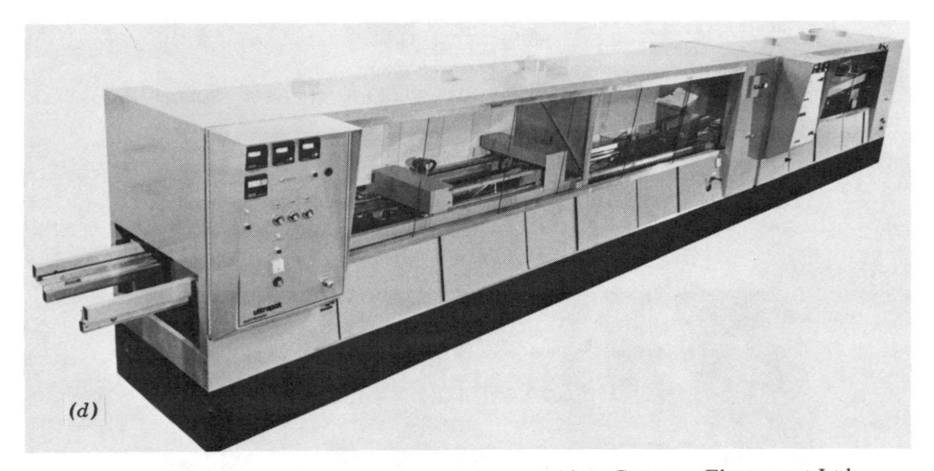

Figure 8.1. (*d*) A complete solder–cut–solder machine. Courtesy Electrovert Ltd.

The difficulty, of course, is to provide a soldering machine that will allow the passage of boards with long leads through the fluxing and soldering stations. The problem is overcome in several ways, each with its own advantages and disadvantages. Any of the methods of soldering using the static solder bath described earlier in this book will allow the soldering of these long leads. Where the leads are less than 1 in. (2.5 cm), there are special deep waves available that can be used. The static solder baths almost always involve more complex conveyor mechanisms to move the assemblies smoothly through the solder, and to achieve the correct entrance and exit angles. In other systems, the solder bath is moved in synchronism with the movement of the board. Whichever is chosen there is an added complexity to the mechanics of the machine.

The deep solder wave, on the other hand, is limited to the length of lead that can be processed, and the higher, and therefore more turbulent, wave inevitably produces more dross. In spite of this, it is a good choice when the lead lengths can be kept within the limits that can be accepted by the machine, because of the simpler mechanics.

While pointing out the problems encountered with these systems, this is not to infer that any of these methods of soldering are not good practical processes. Certainly they should be avoided if possible because of their added complexity, but for specific purposes they offer an alternative soldering and assembly method which may prove to be economically sound. All are in use and all are capable of producing excellent results. After the assembly is soldered it is passed through a high-speed rotary cutter where all the leads are cut to length. The product is then run through a second soldering stage which consists of a perfectly standard soldering machine; hence, the name solder-cutsolder, or SCS.

The second soldering station is primarily to solder the cut ends of the component leads, so as to comply with certain specifications, to remove any shorts or bridges caused by the first soldering, and to rectify any cracks or stresses caused by the cutting stage. At least these are the reasons normally given for the second soldering. In fact, many companies have found that where the product requirements do not require the cut leads to be covered with solder, they can delete the second soldering. There are adequate data in the form of user experience to show that a properly designed, well-sharpened rotary cutter will not damage the solder joint or affect its long-term reliability in any way. Tests in the laboratory with an accelerometer placed on the joint have produced figures that indicate that the stresses produced by this form of cutting are, in fact, less than those developed by manual cutting, especially if the hand tools are not correctly used or are not sharp. Where the boards are not complex, where the circuit spacing is not excessively small, and where the product specification permits, it is well worth the time to carry out tests to see if the first soldering station can provide the level of solder joint quality that is necessary and economically acceptable.

THE WAX STABILIZATION SYSTEM (Fig. 8.2)

This method of holding the components is in many respects similar to the solder-cut-solder, except that a wax is used to lock the leads for cutting, rather than the solder. This wax is especially formulated to perform a dual function: it locks in the leads during the lead cutting, and is then melted to act as a flux for the soldering operation. In the machine the wax is melted in a pot that is, for all intents and purposes, a solder pot, with a similar pumping system that produces a very deep wave. The board is transported over this wave where the wax flows over the board, up into the holes, and on cooling effectively holds the components in position. The cooling can require a separate stage with blowers forcing either cold, ambient air or in

Figure 8.2. The wax stabilization system. (*a*) A complete machine.

some cases refrigerated air over the assembly to ensure complete freezing of the wax and effective lead retention. The wax is normally heated to 140°F (60°C) and the wave will accept component leads up to 1.6 in. (4 cm) in length. After cutting the leads the assembly goes to the soldering station, which is virtually a standard soldering machine, except that rather more preheat is necessary, and of course there is no need for a fluxing station, as the wax provides all the fluxing action necessary. The preheat has to melt the wax so as to leave only that which is required to produce a sound joint, while at the same time raising the joint to a sufficiently high temperature so that soldering can proceed at the required speed.

This method of retaining the components is somewhat messier than the SCS, and, of course, has to contain the additional cooling module. It is claimed that the boards are not subject to such large thermal stresses as with the SCS as the wax is not heated to the same temperature as the solder; of course, the use of a wave instead of a static bath maintains the mechanical simplicity of the machine, without the generation of dross, and without the limitations in lead length. It does not matter if the wax wave is turbulent, or if excessive wax is put onto the board, as it is eventually melted off. It is also claimed that the cost of the wax is lower than that of the solder.

Against this some users have reported fluxing problems with the use of the wax as a dual function material, and some instances of uncertain retention during lead cutting. This is a complex issue, and much depends on the volume and type of product that is to be run. If the use of an in-line lead

Figure 8.2. (*b*) A close up of the wave of wax used to lock in the component leads. Courtesy Hollis Engineering.

locking and cutting system will improve your productivity, make some tests on all of the systems that are available:

Run a reasonably sized sample, at least 20 assemblies on each machine.

Use the most complex board for the tests.

Kovar or other iron based leads are the most difficult to cut cleanly, so use an assembly with DIPs or similar devices with these lead materials.

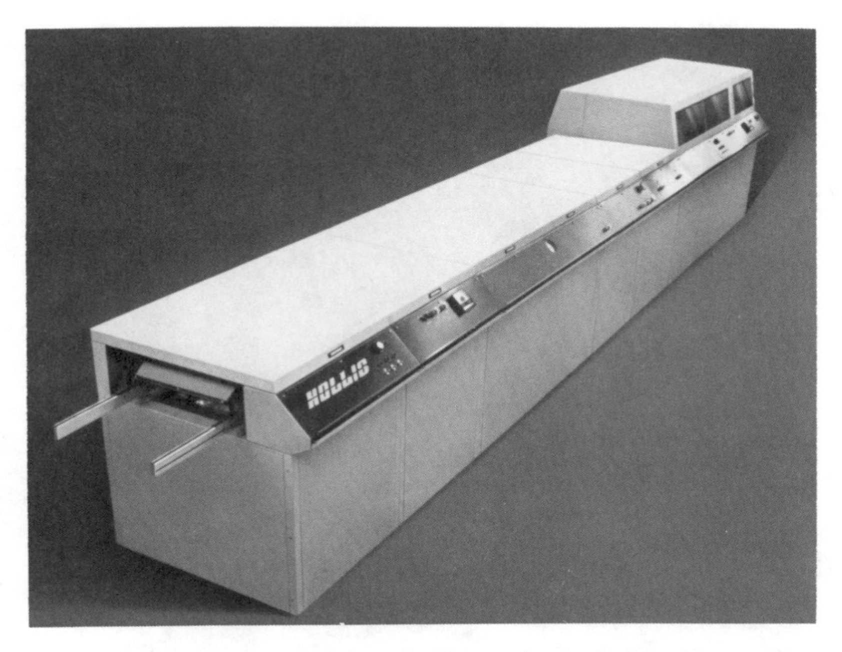

Figure 8.3. This illustration shows a system which can be obtained as either a solder–cut–solder machine or as a wax stabilization system. Courtesy Hollis Engineering.

Look at the maximum throughput attainable with each system.
Check the results, the cost, and the quality.
Use the results to determine your choice.

These systems are continuously being improved and updated to improve their throughput and capabilities (Fig. 8.3).

Chapter 9

The Future of Machine Soldering

Machine soldering has been around as the main method of making the electrical connections on PWBs for over 30 years, and there seems to be no reason to think that there is any other process that will supplant it in the near future. However, it is certain that this process is at the point of considerable change, and that other forms of mass interconnection will intrude more and more onto the scene.

The changes in the machine soldering process have already been mentioned in the main text of this book: greater control of all the parameters of the machines, to much tighter tolerances than were formerly considered essential. Computer control of the system is becoming a common feature, with the operator only monitoring the output of the machine. Initially, these machines are controlling the parameters that have been experimentally determined and placed into the computer memory. Eventually sensors will be developed so that the computer will be able to assess the variables in the process, and make adjustments to the machine in real time, depending on these inputs. The systems will be able to measure the thermal characteristics of the assemblies, check the solderability of the parts, and continuously modulate the parameters for optimum results. The machines will perform their own maintenance, continuously analyze the solder and flux, and make their own additions, or dump and refill.

Machines are becoming much smaller in spite of the increase in complexity, and are also more energy efficient. As well as a reduction in power consumption, more effort will be made to reduce the emission of contaminants into the atmosphere, without the need for the present-day venting systems.

The soldering machines will remove many of the variables from the process, and permit the soldering of circuits which today would be considered much too tightly packaged.

As discussed in Chapter 7, the emphasis will move to the requirements for excellence in the solderability of both components and PWBs. For both of these, only hot solder coating will be acceptable in the near future, a trend that is already beginning. Hot air leveling of PWBs, because of the inherent quality testing incorporated into the process, and a similar method of applying solder to the component leads will be used. Solderability testing will become as important and as commonplace as electrical testing is today.

While it is unlikely that the actual solder will change much in composition, except perhaps to see more emphasis on purity, there will almost certainly be major changes in the ingredients used in fluxes. Already on the market are fluxes that are said to be totally inert after soldering, both electrically and chemically, and do not need to be removed. This may be one way that the flux chemists can reduce the cost of board cleaning. There will certainly be an increasing use of water soluble fluxes, and more types will be developed.

The use of the simple chip components, mounted on the solder side of the PWB, is already widespread, and this will doubtless increase for some years to come. It offers a simple method of increasing the packaging density without going to the expense and complexity of the multilayer board. The soldering of these components has been satisfactorily carried out on the normal soldering machine with the help of the newly developed nozzles.

Of course, it would be foolish not to acknowledge the fact that more and more of the interconnections are being made in the semiconductor chip itself. For many years this was said to forecast the demise of the PWB but the many forecasts were proved wrong, and the volume of boards produced and soldered has increased at a phenomenal rate. There is little reason to suppose that this trend will not continue, but the increasing use of surface mounted components may require changes to the methods that are used to mount and solder these parts. It is not likely that they will be used to any great extent for some time, except in the fields of military and aerospace, where cost is not a factor, until a suitable form of making mass connections can be developed.

The soldering machine can make many thousands of reliable joints at a very low cost, and this must remain an economic factor in the development of any new packaging developments. This is not to say that some new interconnecting method may not be invented which in turn may spawn a totally new form of electronic packaging in the same way that the mass soldering process went hand in hand with the development of the printed wiring board.

At the moment however, there is no indication that this is about to happen, so there is every possibility that soldering will provide the electrical interconnections for many years to come—a much more controlled, refined, and sophisticated process that is generally found today, but fundamentally not much different from that which has served the industry well for many years. Growing up in a parallel with this increasing sophistication will be other soldering processes, capable of very high speed surface mounting and soldering of mutlileaded components, capable of producing joints of the highest quality, consistently and automatically.

There will also be a definite upgrading of the skills of the people involved in practicing the soldering process. As it turns from a black art to a science and from a simple rather crude machine to a computer controlled, precise process, the people will also move from a loader and unloader of assemblies to a highly trained technician responsible for the most critical operation in the electronics industry.

Aside from these crystal ball glimpses into the future, there is one more change that will be seen that is no dream but totally factual. Cost alone, cost of inspection, cost of touch-up, and the cost of a company's reputation for quality and reliability will force more and more organizations to review their soldering methods and organization. This will then turn them to the philosophy of zero defect soldering as the only way to resolve their soldering quality problems.

POOR QUALITY SOLDERING IS THE MOST EXPENSIVE PROCESS IN THE ELECTRONICS INDUSTRY.

ZERO DEFECT SOLDERING IS THE LOWEST-COST METHOD OF MAKING ELECTRICAL INTERCONNECTIONS.

Chapter 10

Useful Information

It is impossible for any soldering specialist to have a detailed knowledge of all the many disciplines that make up the soldering process. It is, therefore, extremely useful to be able to turn to the specialists in any particular field for assistance.

The lists of companies and organizations given here are to help in finding this assistance. This is by no means a complete list, and many fine sources will inevitably be omitted.

The presence or absence of any address in this list must not be construed as any form of comment on the organization or company concerned, or its capabilities. These are sources of assistance that 1 have known during my years in the industry, who may in turn be able to help the reader.

SPECIFICATIONS AND STANDARDS

The Institute for Interconnecting and Packaging
Electronic Circuits (usually called the ICP)
7380 North Lincoln Avenue
Lincolnwood, IL 60646
Tel. (312) 677-9570.

The IPC will supply on request a complete listing of all the available specifications and standards, but the list that follows identifies those which may be particularly useful.

IPC-T-50B	Terms and Definitions
IPC-D-320A	End Product Specification for Single- and Double-Sided Printed Boards
IPC-A-600C	Acceptability of Printed Wiring Boards (A compilation of Visual Quality Acceptability Standards)

IPC-CM-770B	Guidelines for Printed Wiring Board Component Mounting
IPC-S-804	Solderability Test Methods for Printed Wiring Boards
IPC-S-815A	General Requirements for Soldering of Electrical Connections and Printed Wiring Assemblies

The IPC Assembly-Joining Handbook is especially recommended as an excellent guide for the various techniques involved in component mounting, including the design parameters, as well as the tools and methods.

The IPC also publishes the papers presented at the semiannual meetings, together with audiovisual materials covering many of the processes involved in the fabrication and assembly of electronic products.

American National Standards, usually referred to as ANSI documents can be obtained from
ANSI (American National Standards Institute)
1430 Broadway
New York, NY 10018

British Standards can be obtained from
British Standards Institution
Sales Office
Maylands Avenue
Hemel Hempstead
Herts. HP2 4SQ, England

The following organizations will supply standards, or information on obtaining standards for their respective countries.

Institut Belge de Normalisation
29 Avenue de la Brabanconne
B-1040, Bruxelles 4, Belgium

Association Francais de Normalisation (AFNOR)
Tour Europe, Cedex 7
92080 Paris La Defense, France

Deutscher Normenausschuss
4–7 Burggrafenstrasse 1
Berlin 30, Federal Republic of Germany

Ente Nazionale Italiano de Unificazione
Piazz Armando Diaz 2
1 20123 Milano, Italy

Nederlands Normalisatie-Instituut
Polakweg 5
Rijswijk (ZH) 2108
The Netherlands

United States Military Specifications can be obtained from

Naval Publications and Forms Center
5801 Tabor Avenue
Philadelphia, PA 19120

From the many hundreds of Mil Specs and Stds available the following list covers those of most interest to those involved in machine soldering.

Mil Std 202E	Test methods for electronic and electrical component parts
Mil P 28809	Printed wiring assemblies
Mil Std 275	Printed wiring for electronic equipment
Mil P 55110	Printed wiring boards
Mil F 14256	Flux, soldering, liquid (rosin base)
Mil P 81728	Plating tin-lead
Mil C 14550	Copper plating

SOLDER MACHINE MANUFACTURERS

This list is by no means comprehensive, but covers the majority of the more well-known manufacturers in the United States, Canada, Europe, and Japan.

Europe

Light Soldering Developments Ltd.
97 99 Gloucester Rd.
Croydon, Surrey CR0 2DN, England
Tel. 01 689 0574

Roken Electronics (Europe) Ltd.
Unit 3E West Way
Walworth Industrial Estate
Andover, Hants., England
Tel. Andover 0264 57377

E.P.M. Handels A.G.
Ch-8035 Zurich
Beckenhofstrasse 16
Switzerland

Zevatron GMBH.
Postfach 1220, D 3548 Arolsen I
Federal Republic of Germany
Tel. (05691)668

Japan

Tamura Seisukusho Co. Ltd.
No. 19–43 1-Chome Higashi Oizumi-machi
Nerima-ku
Tokyo, Japan 177
Tel. (03) 925–1111

Nihon Den-Netsu Keiki Co. Ltd.
27–1 Shimomaruko, 2-Chome
Ohta-ku, Tokyo Japan
Tel. (03) 735–1231

Koki Co. Ltd.
4th Floor, No. 2 Marutaka Bldg.
7–13–8 Ginza
Chao-ku, Tokyo, Japan 104

North America

CycloTronics Inc.
3858 N. Cicero
Chicago, IL 60641
Tel. (312) 282–6141

Electrovert Ltd.
3285 Cavendish Blvd.
Montreal, Quebec H4B 2L9, Canada
Tel. (514) 488-2521

Hollis Engineering
15 Charron Avenue
Nashua, NH 03063
Tel. (603) 889–1121

Sensby USA
2131 19th Avenue
P.O. Box 16901
San Francisco, CA 94116
Tel. (415) 421-4727

Unit Design
1140E Valencia Drive
Fullerton, CA 92631
Tel. (714) 526-0800

Technical Devices Co.
11250 Playa Court
Culver City, CA 90230
Tel. (213) 870–3751

The John Treiber Co.
11233 Condor
Fountain Valley CA 92708
Tel. (714) 557–1821

SOLDER AND FLUX VENDORS IN THE UNITED STATES

As well as selling these materials, they will carry out solder analyses, and offer assistance whenever soldering problems arise, or new requirements require a change in the chemistry of the process. This is by no means a complete list of all the vendors.

Alpha Metals Inc.
600 Route 440
Jersey City, NJ 07304
Tel. (201)434–6778

Frys Metals Inc.
6th St. and 41st St.
Altoona, PA 16602
Tel. (814) 946–1611

Gardiner Solder Co.
4820 S. Campbell Avenue
Chicago, IL 60632
Tel. (312) 847–0100

Kenco Alloy and Chemical Co Inc.
418 W. Belden Avenue
Addison, IL 60101
Tel. (312) 543–6060

Kester Solder Division
Litton Systems Inc.
4201 W. Wrightwood Avenue
Chicago, IL 60639
Tel. (312) 235–1600

Multicore Solders
Cantiague Rock Road
Westbury, NY 11590
Tel. (516) 334–7997

Temperature indicating crayons can be obtained from

Tempil Div.
2901 Hamilton Blvd.
S. Plainfield, NJ 07080
Tel. (201) 757-8300

Some other contacts that the author has found useful. This company manufactures the Edge Dip solderability tester.

Williams Machine
2092 W. Main Street
Norristown, PA 19403
Tel. (215) 539-2123

Quartz heaters of various sizes and types can be obtained from

Casso-Solar Corp.
3 Pearl Court. Allendale Industrial Park
Allendale. NJ 07401
Tel. (201) 825–4600

Flux density control systems can be obtained from

Norcross Corp.
255 Newtonville Avenue
Newton, MA 02158

Celmacs Corp.
25067 Viking Street
Hayward, CA 94545
Tel. (415) 785-3390

Sensby VSA
2131 19th Av.
P.O. Box 16901
San Francisco, CA 94116
Tel. (415) 421-4727

Solder joint inspection machines are made by

Vanzetti Systems
111 Island Street
Stoughton, MA 02072
Tel. (617) 828-4650

Benchmark Industries Inc.
P.O. Box 3160
Manchester, NH 03105

It is sometimes necessary to look for outside help in solving processing problems, especially when they involve disciplines outside the normal scope of an electronics assembly house. The first line of attack should be through the vendors of the materials involved, who will have expertise in the appropriate fields. In case of emergency a local university can often be used to advantage; often another company can be persuaded to provide assistance on a consulting basis. Membership of one of the trade organizations can often lead to useful contacts. Some other sources of technical assistance are:

The Tin Research Institute Inc.
2600 El Camino Real
Suite 224
Palo Alto, CA 94306
Tel. (415) 327–6650

or

483 West 6th Avenue
Columbus, OH 43201
Tel. (614) 424–6200

Lead Industries Association Inc.
292 Madison Avenue
New York, NY 10017

Trace Laboratories (who offer independent testing services)
9030 Eton Avenue
Canoga Park, CA
Tel. (213) 341–5593

or

P.O. Box 8644
Baltimore/Washington Int. Airport
Baltimore, MD 21204
Tel. (301) 859–8110

WoodCorp Inc.
Shore Drive RR#4
Brewster, NY 10509
Tel. (914) 279-2899.

Training is a specialized business, and when it is not possible to find in-house facilities, the following companies may prove useful.

Omni Training Corp.
20620 Arrow Highway, Unit 5,
Covina, CA 91724
Tel. (213) 967–0727

Tatham Training Corp.
One Yonge Street, Suite 1502
Toronto, Canada M5E 1E5
Tel. (4I6) 367–1870

Pace Inc.
9893 Bravers Court
Laurel, MD 20810
Tel. (301) 490–9860

WoodCorp Inc.
Shore Drive RR#4
Brewster, NY 10509
Tel. (914) 279-2899.

Tempered glass panels for setting up the solder wave can be obtained from

The Hexacon Electric Company
161 W. Clay Avenue
Roselle Park, NJ 07204
Tel. (201) 245–6200

Index